I0036421

SIMPLIFYING MINING MAINTENANCE

Simplifying Mining Maintenance

A PRACTICAL GUIDE TO BUILDING A CULTURE THAT PREVENTS BREAKDOWNS AND INCREASES PROFITS

Gerard Wood

LIONCREST
PUBLISHING

COPYRIGHT © 2018 GERARD WOOD
All rights reserved.

SIMPLIFYING MINING MAINTENANCE
A Practical Guide to Building a Culture that
Prevents Breakdowns and Increases Profits

ISBN   978-1-5445-1253-2 *Paperback*
       978-1-5445-1254-9 *Ebook*

I would like to dedicate this book to my late uncle Gordon Wood, who lost his life on a mine site where I worked at the time. His death is always a reminder of the hazards that we must look out for and control every day in mining, and why it is necessary to take the management of safety so seriously.

Contents

Acknowledgments

I would like to acknowledge all of the people whom I have worked with over my career. This book is really a compilation of the experiences and learning that I have taken from working with so many talented people over the past thirty-four years.

Introduction

MY 360-DEGREE JOURNEY

I started my career on the workshop floor in the mid-1980s, maintaining equipment as a tradesperson. In Australia, where I live and work, we call these folks "tradies"—and I always enjoyed being one. I took a lot of pride in my work, and I felt responsible for keeping my machines running reliably without any unexpected breakdowns. You never wanted to come into work and learn from the night shift that one of your machines had broken down. That made you feel terrible, so you did everything you could to prevent that from happening. I did not want to be the guy who caused problems for the other shifts. No one did. It made you feel like you'd let a coworker down.

Over the next fifteen years, I moved up from the shop floor into a supervisory role, then into planning, engineering, and maintenance management roles. While working full time in the mines and growing my wonderful family, I studied electrical engineering, mining engineering, and business administration.

Eventually, I moved into a central maintenance role, and my job was to help many different mining sites along the east coast of Australia and other parts of the globe improve their maintenance efforts. I was on a team—Global Maintenance Network (GMN)—that worked with the local mines to improve the reliability of their machines, which exposed me to models, theories, and frameworks for equipment maintenance, reliability, and asset management. We used to joke that in the GMN we had more models than a hobby shop. All of these models were flawed, but some were useful. We mapped workflow processes, conducted maintenance evaluations, did Six Sigma process improvement, and conducted root-cause analysis, among many other improvement initiatives.

But even after several years of this, I didn't see much improvement in the reliability of the mining machinery. We rolled out models, trained staff, taught theories, and invested in computer maintenance management equipment, but we never saw the results we should have. The equipment should have grown increasingly reliable, but it didn't. Of course, there were pockets of improvement, but generally speaking, there was no companywide consistent improvement in equipment performance.

I realised after some time that we had overcomplicated the process of maintenance so much that the fundamental value of putting a person with the right level of ownership to work on the machines—a machine the mechanic knows personally and has worked on extensively already—had been lost. **Processes and systems are essential, especially in large businesses, but without the basics of the work being executed well, all of the benefits from the processes and systems are lost.** With this realisation, my approach to maintenance had come full circle. I began thinking again like a tradesman, with the understanding that caring for the equipment and maintaining it in good condition was the best way to prevent failures.

I have since worked extensively with companies to rebuild that culture of ownership and quality and instil confidence that maintenance can make a difference. These companies have achieved great results in a short time.

This success inspired me to take on the daunting task of writing this book. My goal in writing it is to make that culture of ownership and quality maintenance commonplace in our industry. I want the mining industry to be seen as **leaders in asset management and equipment reliability.** I want to remind people how important crafts-manship is and show them how they can incorporate it back into even the most complex maintenance strategies. I want this book to bring the focus back to the basics.

A CAUTIONARY TALE

When it comes to mining, equipment maintenance is not an option or a luxury. The work is too hard on the machines. When you are using equipment to haul tons of ore or bore into bedrock for hours a day, proper preven-tive and proactive maintenance is a necessity.

Many mining companies have spent millions and some-times billions of dollars to improve maintenance systems and equipment reliability only to discover little return. Senior managers continue to complain about equipment reliability and maintenance costs.

It doesn't have to be this way. This book is not meant to be a detailed manual on how to technically implement all these little systems to treat the symptoms. Instead, the book's goal is to help people in the mining industry agree on the importance of maintenance. I'm not advocating a technical approach. In fact, I recommend keeping things **simple**. Schedule the required maintenance downtime and do the work when you're supposed to. Give people the time and tools they need and let them *prevent* problems rather than simply attacking them when they crop up.

You don't need to read this book sequentially; simply read the appropriate chapters when you need ideas for dealing with your problems. If you have difficulties with only handover to operations, then skip to that chapter. If you spend too much on maintenance, read the section on cost problems. Focus on the areas where you are struggling. Use this book as a guide. It is not a 100 percent solution, but it will offer ideas for practical results. The methods I recommend will free you from the mountains of checklists and procedures that tie up their mechanics and administrators and keep them from more important work.

I want mining maintenance people to believe they can be the best-performing maintenance teams in the world. I want them to believe in their value and to rekindle their passion for improvement. I want people to simplify complicated processes and documentation. I want to make it easier for people to get their work done and hold one another accountable. I also want to give them the tools, space, and consistency they need to rediscover a sense of pride and ownership in their work. That pride and workmanship is what keeps equipment running and makes a mine successful. I want to encourage that culture.

My ideas on simplifying and improving mining equipment maintenance have been influenced by a number of leaders in the mining industry. I have learned from everyone I have worked with over the past thirty-three years, and many of these people gave me critical insights into simplifying and improving the industry.

When I worked for BHP and was in the GMN, we purchased a company called Western Mining Corporation. A guy named Dick Pettigrew was their head of reliability. While those of us in the GMN focused on evaluations, fluid cleanliness, root-cause analysis, and defect elimination, Dick concentrated on a whole-of-business approach to reliability improvement. Dick considered reliability-centred maintenance (RCM) as the fundamental framework for building reliability. Dick came from the petrochemical industry, and even though RCM hadn't been very successful in the mining industry, Dick showed me that RCM could help solve our reliability problems—not as a project (or something we jokingly called a Resource Consuming Monster) but as a way of thinking about managing equipment failure modes to improve reliability. RCM allows you to accurately manage specific failures so you can avoid them in the future.

Dick also showed me some frameworks around operational discipline, convincing me that we cannot progress if we don't do what we say we will do and follow simple procedures. We get stuck in a loop (below) that does not deliver sustainable improvement. Without first verifying that we are following current processes, standards, and procedures, we cannot advance.

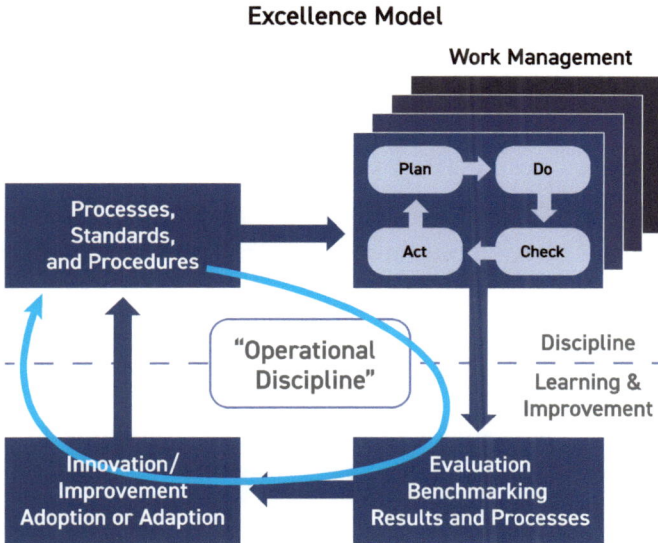

Excellence Model

Work Management

Plan

Do

Act

Check

Processes, Standards, and Procedures

"Operational Discipline"

Discipline

Learning & Improvement

Innovation/ Improvement Adoption or Adaption

Evaluation Benchmarking Results and Processes

Ian Goodwin, my former boss in the GMN, taught me the value of engaging people to ensure they take responsibility and feel pride and a sense of ownership in their work. When you're an engineer, it's easy to focus on the technical aspects and forget about the importance of people, but from Ian, I learned how to make sure mine employees and managers—and not just the executives—develop solutions. With this approach, workers own and drive any changes in procedures or practices. Ian is also a master at networking and knowing who will have the answers to specific problems.

I also worked at the GMN with Phil MacMahon. Phil was an engineer, and he didn't come up from the workshop floor the way I did. Nevertheless, he once wrote a paper titled, "How We Killed the Craftsman," and presented it at one of our conferences. Phil understood the technical theories and had been rolling them out for years before I started, but he also recognised that in mining maintenance, we had killed off the passion, pride of workmanship, and desire among tradespeople to do a good job. His paper hit the nail on the head. He understood that ownership and passion among tradies was essential to improving equipment reliability. When Phil became maintenance and operations manager at another company, he achieved great results with systems that allowed tradespeople to do their jobs right. Phil did not have the same journey as me, but we both ended up in the same spot.

Another colleague, Rod Bennett, was an expert at detecting potential defects. By analysing vibration and temperature, Rod saw signs of failure far sooner than many people in his field.

Rod concluded that if people do maintenance correctly—conducting proper proactive maintenance, installing the right components properly, and consistently doing quality work—machines run reliably and defects are eliminated. You didn't have those unscheduled and costly breakdowns or early-life failures that cause so many problems for the business. "Machines don't die," he once wrote. "They're murdered." He, too, believed that we maintain mining equipment with little regard for quality or what he called "the basics." Mine managers need to have people who know how to do alignments, fasten bolts correctly, keep lubrication clean, and do a proper weld. Rod understood that if mine managers did all those things properly, the machines would be more reliable and less costly to maintain.

The work of these men opened my eyes to the core aspects required for good equipment outcomes, an understanding of how to manage failure modes effectively, and the importance of ownership and quality maintenance work. The theories, computer systems, and frameworks are essential but also useless without this practical foundation. It seems strange, but mining executives or maintenance leaders can walk around a mine site and look at the equipment, evaluate the maintenance program, and talk to the people on the floor and still miss the problems. They don't see the leaks around a hydraulic hose coupling, the amateurish weld, or the ominous high temperatures from faulty cooling systems. If you haven't trained your eye to look for these things, you miss them.

When I visit a mine site, I like to go out into the field with the maintenance leaders, look at these little details, and ask why these conditions exist given that we generally have a maintenance program to prevent them. These are the things that cause machines to fail, but until we focus on them, we remain blind to the fact that they are causing our problems.

Rod Bennett also introduced me to the concept of a reliability visit or reliability walk. This is where you go out into the plant and look specifically for issues affecting reliability. An example format for this is provided in chapter 2. Being able to go into the plant and critically assess conditions that lead to reliability is an essential organisational capability to develop.

HOW TO BECOME PROACTIVE, NOT REACTIVE

People in a reactive mining maintenance environment are busy. The phone constantly rings during the workday, and they get calls every night and on the weekends. The worst part about it is that they deal with the same problems over and over. They never have time to plan because they run from one emergency breakdown to the next.

I hope this book will help change that. In my experience, well-run maintenance organisations *do* have time. When everything is scheduled and goes according to plan, these departments don't have all that drama.

The tools I present here should help them transition from a reactive workplace to a proactive workplace where planning and scheduling deliver the intended business improvements. I hope to destress their professional lives and give maintenance managers back the time they need to do their jobs with quality, precision, and forethought. I want them to enjoy their weekends again.

I understand the critical role of maintenance. I've worked all over the world for various mining companies at all different levels, and this experience has allowed me to understand what the general manager (GM) and the chief operating officer (COO) want to see. They want to see more profit! It's not complicated.

I also realise what the maintenance manager is going through and how his relationship with the GM and COO can be damaged. When equipment becomes unreliable, the maintenance manager usually says he needs more people and money to correct the situation. This is not what the GM and COO want to hear.

I hope maintenance managers will use this book to find

practical solutions to common dilemmas we all face. Every time I help people at a mine site, I see the same struggles over and over—difficulties I've seen solved many times before. There is no reason why these problems can't be eliminated so maintenance people can get back to performing proper and precise preventive maintenance that keeps their machines running well. In part I of the book, I describe the experiences and mistakes that taught me about the major pain points, and in part II, I discuss how to solve some of these pain points in a simple manner.

I also hope COOs and other executives will use this book to gain insight into how to improve maintenance and reliability. They'll learn how to judge maintenance. They'll come to understand why they must look at maintenance in the long term. If their expenditures are for planned and scheduled maintenance, executives should expect fluctuations in expenditures from one year to the next. That doesn't mean everything is bad in the maintenance area.

I'll also help executives understand the difference between scheduled and unscheduled service, and the metrics they need to better understand how this preventive work is performing. Executives understand the value of servicing their personal cars. Preventive upkeep on mining equipment is the same; it protects your investments and pays dividends for the business. Keeping this in mind helps us simplify our approach to mining equipment maintenance and operations.

I'll also introduce a couple of models that I like to use. I know, I know. I just told you how we've come to rely too much on complicated models and theories. However, these models don't require companies to restructure their operation. Instead, executives and maintenance managers can use these models to change and improve maintenance. The models establish the foundational values for maintenance improvements and provide a way of thinking. They don't prescribe procedures.

Finally, I want leaders—whether they are a COO or a maintenance manager—to take extreme ownership of the maintenance outcomes. The buck stops with you.

People often like to blame others for equipment breakdowns. They blame the manufacturer, or they blame the contractor. But maintenance managers and GMs have to accept some responsibility for equipment reliability—such as when the maintenance manager pulls people off scheduled maintenance so they can repair an emergency breakdown, or when the GM cancels scheduled maintenance to keep a machine working. These people must understand these actions contribute to the poor reliability of their equipment.

Writing this book made me reflect on my first maintenance manager's role. I was not getting the results I wanted from two fleets. I started to blame the guys who supplied and overhauled our components because some had failed, but then I realised that I had selected the suppliers. I had to own that decision with zero excuses.

If you don't have that mindset—that level of extreme ownership and zero excuses—then you won't take steps to correct the problem. The result is a mine that does not produce—something I've seen happen a number of times—or a mine that isn't nearly as profitable as it could be. If you want to be part of a highly productive and profitable mine, then read on to learn the simple steps you can take to achieve it.

I'm not suggesting my maintenance operations always ran perfectly. We often had excellent performance, but my maintenance teams were never perfect. That's why I can relate to people going through the same experiences. Despite occasional setbacks, I focused on continually improving the equipment reliability and eliminating breakdowns.

I don't have all the answers, but I do know one thing: we can control all failure modes. What worked for me can work for you.

Are you ready to take control?

PART I

An Asset Management Journey

CHAPTER 1

Experiencing All Roles

My first job in mining was as an apprentice electrician on the shop floor at a BHP coal mine, working exclusively on preventive maintenance and rarely on emergency repairs or breakdowns. I was studying electrical engineering at night to earn my degree.

At the time, computerised maintenance management systems (CMMS) were just emerging and had not been used in mining, but our team still achieved good reliability with the machines we serviced. CMMS are designed to help you keep track of a complex machine's or large fleet's scheduled and corrective maintenance activities, lubrication, alignment, and the like, but we relied more on what we saw and heard in our inspections. We listened for unnatural vibrations, inspected for worn parts, and looked for any signs of overheating or lubrication leaks. Most of us had a strong work ethic and a genuine desire to make sure these machines never broke down. At that time, each tradesperson had areas of the machines that they were responsible for, so if that area suffered a breakdown, everyone knew who missed the signs of the defect. You didn't want to be that guy.

Consequently, most tradespeople had a high level of ownership and pride in their work. They didn't need someone looking over their shoulder to make sure they did the job right. The tradies felt responsible when a machine broke down, so they worked hard to prevent that from happening. Their motto was, "Do it once, do it right the first time."

The mining industry started using the CMMS about the time I became a supervisor. The CMMS helped us to plan maintenance and improve our machines' performance by enhancing the efficiency of our scheduled maintenance and keeping a record of all the machines' defects. This was successful because the core maintenance work was already done correctly, so the availability of our machines improved. Many of us thought the CMMS was the solution to all our problems.

Over time, however, the computers began to drive the way we worked. Before the CMMS, we made up for a lack of maintenance planning with workmanship and attention to detail. After the computers took over, that attention to detail eroded as the mining industry went through workforce changes and the computers dictated when and where we should work. We lost the deep knowledge of our machines' conditions. We stopped listening for ominous vibrations or looking for suspicious leaks because the computer didn't tell us to.

Over time, we just measured whether the job was opened and closed, not how the work was executed. These days, we try to fix work-execution quality problems that lead to unscheduled maintenance events with planning and scheduling through CMMS systems. Unfortunately, planning and scheduling creates more efficient scheduled downtime but not improved reliability.

BACK TO THE BASICS

I worked as an apprentice for four years, then as a tradesperson for four or five years before becoming a frontline supervisor. After several years as an electrical and mechanical supervisor, I went on to become a maintenance planner, a superintendent, an engineer, and then a maintenance manager before I joined the Global Maintenance Network (GMN). At this time, I was exposed to the theories and the term "asset management," which we used to call maintenance and reliability.

At GMN, I was on a team that rolled out the training in work-management processes, including defect elimination, root-cause analysis, Six Sigma, the SAP CMMS, and so forth. We introduced these workflow processes to various sites, but it was up to the people at those sites to implement them.

What I noticed over the next five years was that we didn't get the reliability we expected. I visited these sites where we'd done all this training and implemented these programs, and the results were poor or no better than before. Business didn't improve, and equipment didn't become more reliable. Consequently, the company revised all the maintenance work management procedures. The managers figured workers didn't follow the process because they didn't understand them, so they wrote them in greater detail. Unfortunately, the procedures became more complicated and even fewer people understood them.

This frustrated me, so I started studying reliability-centred maintenance and other alternatives. I looked for solutions that could be implemented across multiple sites that would make a company-wide difference.

I also decided to go back to the line-management game. I wanted to reconnect with my roots and go back to the way I was taught—the practical side of things. Instead of traveling around from one site to the next and facilitating improvement programs, I wanted to work at the same site every day for a longer period so I could see what it took to improve machine reliability and to implement the theory I had learned. So I took another maintenance manager's position at a coal mine owned by Anglo American and went back to the coal face.

While Anglo was different from other companies I had worked for, the processes were fundamentally the same. However, maintenance processes had become much more complicated than when I was on the floor. It was no longer a simple matter of highly trained tradespeople responsible for a discrete piece of equipment. Instead, tradesmen might work on dozens of different machines, their work changing dramatically from day to day. They were still skilled, but they had no personal history with a machine that would help them spot potential problems and make note of when to address them. The CMMS was good and efficiently deployed people to the right job with the right tools, but these work management metrics were not helping tradies develop the type of craftsmanship that in the past had kept machines running smoothly.

The corporate culture is often to blame when maintenance practices deteriorate. Companies establish key performance indicators (KPIs) for the maintenance crew, and workers stick to those KPIs instead of focusing on quality. Was the job finished? Yes. Was it completed in time? Yes. These guys meet their KPIs, but no one questions the work. No one asks if they noticed any potential problems they should note. There is no KPI that says the guy did a weld that will last or that he cleaned out all the dirt from the component before he fitted it. You can't measure those things, but they are the principal reasons why equipment remains reliable or not.

The workers understand what quality work is, but when no one talks about it or checks to ensure it happens, they also forget about it. They have no sense of ownership because no one expects them to have one.

It became clear to me that we needed to simplify things for all involved. One of the first things I did was implement a structure in which the same people were consistently looking after the same equipment.

I also worked to improve communication. When you're at the same site every day, you can see, for instance, that one planner doesn't like one particular supervisor, and as a result, the two of them never talk to each other. If you want to improve planning and preventive maintenance, these two people must communicate every day, learning from what went wrong and improving procedures so the preventive maintenance can be more efficient.

In addition to these changes, I began coaching my engineers in how to use a simple root-cause analysis process and effective defect elimination. We didn't build a new model, but we tweaked existing processes so they would work better, were easier to carry out, and incorporated the principles of effective management of failure modes. This improved equipment reliability and cost performance.

I learned that tradespeople want their work to be predictable and routine. They want consistency so they can find a rhythm. They don't want things to change all the time. They need repetition in their work so they can get better at their jobs. They get faster, more precise, and can keep on top of the never-ending defects that are continually generated as equipment mines the material in harsh conditions. That's when they take pride in their work because they can see they are making a positive difference and achieving small wins.

STRESSING THE BASICS

This experience and the desire to grow personally inspired me to start my own company, Bluefield Asset Management Specialists. Although we work for many different mining companies, similar to a central maintenance department, our job is more hands-on. Sites engage us to assist them to complete a range of services across the life cycle of equipment—from asset evaluation to operational readiness and reliability improvement.

Our focus is on the practical, simple solutions that bring both short-term and long-term equipment performance improvement. When sites engage us to assist them with their maintenance-improvement effort, we do not go there to fix the problems and leave. Instead, we enable our client's people to work as one team and prevent the problems from continuing. We fade out as the mining company's people take on the challenge.

We don't have a silver bullet for all problems, but the Bluefield Project or Transformation Process that we use engages the hearts and minds of the people and encourages them to take ownership and implement strategies that work. We stress the basics because 80 percent (my guesstimate) of the mining industry does the basics very poorly. There is a tremendous need in the industry for strategies focused on quality execution, ownership, and practical, simple solutions.

These are the strategies I describe in this book. We've had appreciable results with the companies we've worked with over the last seven years, but to improve equipment reliability across the mining industry, I need to share this approach with many, many more people. I can't just keep going out and doing it myself as a service provider or even with our whole team involved; we have only so much time!

I'm writing this book so people can use and tailor this

approach to improve their own operations. Tens of thousands of people can make a difference with the knowledge in this book. I hope you're one of them.

THE VALUE OF THE RIGHT CULTURE

Over the last ten or twelve years, the industry has started referring to maintenance as "asset management," but the terms are not synonymous. Asset management encompasses the entire process of managing an asset—how we buy the asset, how we operate it, how we maintain it, and how we dispose of it. Maintenance is just one part of that life cycle, but it's a significant percentage of the asset management pie.

It's critical for us to keep these definitions in mind because it is so easy to use these terms interchangeably. This book, as the title suggests, is just about simplifying the maintenance aspect of asset management.

In the next chapter, I will introduce my Maintenance Improvement Model, which emphasises the importance of a culture built on a foundation of core values. The model illustrates how to think about continually improving maintenance. It starts with a constant desire to improve. Without a desire to improve and a belief that our actions can significantly impact the performance of the equipment, nothing changes.

When I visit a mine site that achieves excellent results from its equipment, there are two things that always stand out: the people are unflinchingly focused on improving, and they never doubt their ability to influence the performance of their equipment.

At one mine in the Hunter Valley in Australia that I visited recently, the maintenance team didn't have all the latest CMMS, but they had an exemplary attitude. They achieved a nearly 91.8 percent availability rate for their truck fleet, which is outstanding, but all they talked about while I was there was how they wanted to reach 92.8 percent. We visited another site in Ghana that had zero systems in place. The truck fleet was in shambles, but the maintenance crew achieved remarkable performance from their drilling equipment. Drills are notorious for breaking down, but the guy who was in charge of the drills maintained an extreme level of ownership over them. When a mechanic did any work on the drills, the manager wouldn't let the mechanic leave until he had inspected the services performed and had signed off on the quality of the mechanic's work. If an operator damaged a machine, they had to help mechanics fix it before the equipment could be put back into operation.

You must have that kind of devotion to quality. I've never seen a site with reliable equipment that didn't have a crew that was passionate and eager about its work. Maintenance crews can have all the systems and processes and CMMS in the world, but if they don't have a staunch commitment to servicing their machinery, the results will be substandard.

How do you create this culture? Read on to find out. In the next chapter, we'll explore my Maintenance Improvement Model. (I call it my model, but it is a compilation of all I have learned from so many in the industry over my career.) You'll see that a desire to improve is a foundational step that must be supported by the right values required for reaching your goals for equipment reliability and financial outcomes.

CHAPTER 2

Even More Models

This chapter will look at models, but rest assured, I will not be introducing any new complex theoretical frameworks in this book. If you're in mining maintenance, you've probably seen enough of those. This book is meant to be practical, so instead of more convoluted theories, I'm going to describe two simple models that illustrate a way to think about maintenance and work management.

TWO SIMPLE MODELS

The first model is my Maintenance Improvement Model. I created it while doing my MBA. I used it in my first maintenance manager's role to help me and my team agree on what was important to us and to create shared values, goals, and practices.

Initially, I wanted to get something down on paper that would represent all aspects of maintenance and how to improve it. Despite the heaps of models out there, I wanted to create something that was more closely linked to what was important from my perspective.

Many people think all work should revolve around models, but that's not the point of the Maintenance Improvement Model. It demonstrates a way of thinking about maintenance. It's a framework for people to talk about servicing equipment and to establish priorities. No model is the be-all and end-all of maintenance, but this model illustrates the values critical for successfully maintaining your company's assets. For example, many people don't make scheduled maintenance a priority. The Maintenance Improvement Model prioritises scheduled maintenance because more breakdowns result when you don't.

My Maintenance Improvement Model should be integrated into a company's practices. The company and its maintenance workers should embrace the foundational beliefs and practices of this model. The model helps maintenance team members calibrate their thinking, behaviour, and interactions in a common effort to keep machines running reliably. Companies and teams may reword or add their own values, but this model helps mining maintenance teams understand and talk about the challenges they face and how to consistently overcome those challenges.

The Work Management Model is modified from a model I was introduced to years ago called the Pipeline Model. The Pipeline Model focuses on the technical side of maintenance, and the business processes people have to follow. My model simplifies these systems to help teams understand them better.

All companies have a version of the Pipeline Model, and I have observed them continually redefine the same processes, often making them more and more complex and even harder to follow. When a process is too complicated, people are less likely to adhere to it. Companies also try to lock people into these processes using the computerised maintenance management system (CMMS), but unfortunately, the crucial steps that the tradesmen need to take to ensure equipment is dependable can't be dictated in a CMMS and must be carried out on the job. You address problems by getting your hands dirty testing belts and checking hoses for cracks and noticing unnatural vibrations.

Maintenance Improvement Model

Goals
No injuries
No repeat failures
Meet business financial targets

Maintenance Management System

Identify Improvements
Benchmarking,
Breakdown Analysis,
Maintenance Delay Analysis,
Safety Observations,
Area Inspection,
Maintenance Review,
Cost Review/Analysis,
Top 10 Pareto Analysis,
Planned Job Observations

Measures to Prevent Slipping Backward
Document Control,
SWPs,
Maintenance Strategies,
Training,
Culture (Rituals)

Values Foundation/Bricks

Desire and determination to improve	Always consider safety first	Accept responsibility for performance. Face reality.	The right PMs done properly
Scheduled maintenance has the highest priority	Learn from experiences	Motivating people	Development of people and technical skills
Understand equipment in detail	Innovative solutions	Go the extra mile for others and expect the same in return	Open and honest communication

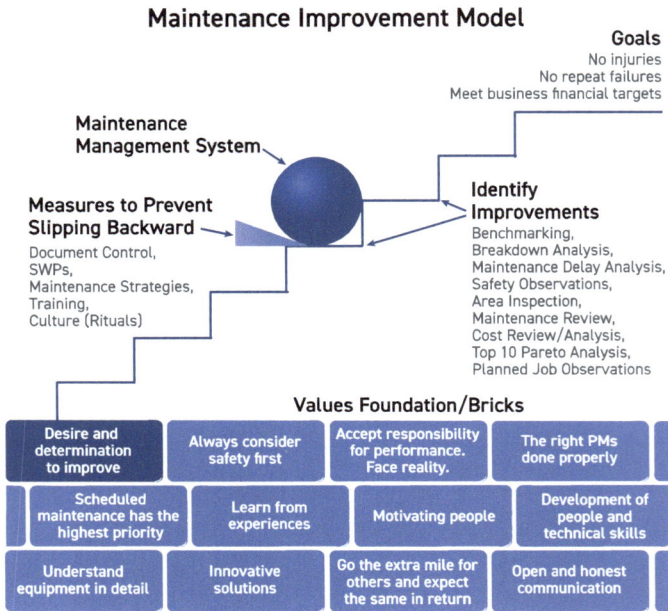

In my model, the big circle depicts the overall maintenance management system. We have to get this maintenance management system ball up the stairs to reach our goal. The goal I have defined here is to have no injuries, to avoid repeat failures, and to meet our business financial targets. You may want to define the goals that suit your business.

THE FOUNDATION VALUES

The Maintenance Improvement Model rests on the foundation blocks—the values at the bottom of the chart. These blocks represent the ways people should think and behave, or the values that create the culture of the organisation. I created this model over the years in maintenance leadership roles, and when I moved into a central improvement role, I added the grey block. Everything that was already in place was great, but I realised that if you don't have that first little step—the **Desire/determination to improve with zero excuses**—you won't make any progress.

In the mining industry, with its staggering investments and monumental risk, maintenance managers must constantly be looking for improvement. The mining companies and maintenance managers who are complacent will not succeed in this industry, because their competitors aren't happy with the status quo. If you don't continually look for your next improvement, you will fall behind.

I've seen this culture of constant improvement many times in mines around the world. Sometimes an entire company will embody this attitude, but I've seen it more often within teams working at particular mines, such as the maintenance team I encountered in the Hunter Valley. They had an embedded philosophy of continual improvement that started from day one.

Once you have that first step in place, you can embrace the other value blocks.

Always consider safety first must be your primary consideration and foremost value. Many companies say that employee welfare is their highest priority, but I view safety as a value rather than a priority. Combing your hair is a priority, but putting your pants on is a value in society. You may wake up late one morning and have to rush off to work without combing your hair, but you won't leave the house without putting your pants on. The welfare of your workers and coworkers needs to be at that level where you don't brush it aside. Priorities can change, but values don't. Keeping your workers out of harm's way needs to be so deeply ingrained in your decision making that you always think about it first.

The next block—**Accept responsibility for performance**—is about facing reality. I read a great book recently called *Extreme Ownership* by Jocko Willink and Leif Babin that epitomised this block.

I had to learn to take ownership of a situation my first time as a maintenance manager. One of our truck fleets had some problems—it was the world's first AC drive diesel-electric truck, and availability slipped below 90 percent. My first thought was, "Well, that's not my fault. The truck has design problems and the equipment manufacturer is letting me down." But I soon realised I had to accept responsibility, because in the end, I had to deal with the breakdowns, and I had the choice to keep using the original equipment manufacturer (OEM) or look for someone who could help me overcome the issues more rapidly. The buck had to stop with me. The moment I accepted this fact, I made the right decisions to eliminate some of the problems using our own team while still progressing the problems that had to be solved by the frustratingly slow OEM.

The next block—**The right PMs done properly**—acknowledges the underlying value of preventive and proactive maintenance, or PM. You must do high-quality work, and it has to be the right kind of work. You must make sure the work and the inspections are done meticulously and that equipment is maintained in a manner that will prevent failures. Mechanical equipment must be maintained in a tight, lubricated, and clean (TLC) condition. Electrical equipment must be maintained in a tight, clean, dry, and cool (TCDC) condition. Many people may also incorporate precision maintenance, error proofing, and quality assurance and quality control (QA/QC). This value block encompasses all of these elements.

The next block states, **Scheduled maintenance gets the highest priority.** In many companies, it is acceptable to take people off planned equipment service to work on a breakdown. This interruption prevents maintainers from finishing service jobs with precision and quality, and that approach puts your operation in a breakdown cycle. Instead, designate people to handle breakdowns and low-priority scheduled work so that important scheduled maintenance isn't delayed or abbreviated. Of course, breakdowns can't be planned; sometimes there are none, and sometimes there are many. When there are more breakdowns than the allocated resources can handle, leave some of the equipment down until workers assigned to breakdowns can get to it. Let workers assigned to scheduled maintenance continue focusing on their work. This approach prevents a costly and chaotic breakdown cycle and will save money in the long run. If you keep your machines clean, lubricated, adjusted, and aligned, you avoid breakdowns. Ultimately, you will have less work to do, which means you will need fewer people and parts.

The next block is **Learn from experience**. On the workshop floor, this often means learning from breakdowns and other unfortunate events. This block serves as a reminder to maintenance personnel that when something doesn't go right, it's critical to note how to prevent it from happening again.

You need to both document it and talk about it. Discussing solutions with everyone on the maintenance team helps them learn, particularly when you can't update written procedures right away. Weekly or daily discussions about specific breakdowns can help people quickly learn how to prevent those breakdowns.

It's impossible to write down every scenario or problem that crops up and even more difficult to educate everyone with written documents, so it is essential that your teams continually learn from experience. Companies that embrace this approach find their teams are more engaged because tradies love to talk about equipment.

The next block, **Motivating people**, may seem obvious, but you must have people who can be motivated to do good work and who will gain satisfaction from more than their salary. As a maintenance manager, it's crucial to create a work environment where people can learn from breakdowns and discuss them openly without feeling threatened or embarrassed. Sharing examples of good maintenance practices motivates those who executed this work. Everyone likes an environment where we have responsibilities and are recognised for an excellent effort.

The next block, **Development of people and technical skills**, reflects the importance of ensuring engineers, shop-floor workers, and everyone else get the technical skills they need. At Bluefield, my company, we have a development plan for each person, and we talk to them about it twice a year. It is part of their performance review, so they know how they should develop and get the career and life skills they need.

Individuals are responsible for their own development, although the company should supply time and financial support for training. Some companies fear losing employees who have received training, but the greater fear should be keeping employees who have not developed.

I like to use a training matrix that encompasses not only formal training but also on-the-job coaching. It is essential to have a deliberate approach to developing all of the necessary technical skills across the team while recognising that not everyone can be an expert in every field.

Developing people leads to the next block, which is **Understand the equipment in detail.** When you get your machines on a site, your mechanics must learn how that equipment works, how to fault-find, and how the equipment fails so that these breakdowns can be prevented. Over time, mechanics will pick up critical nuances, such as which components degrade faster or how dust gets into certain areas and causes premature wear. These tradespeople learn how to prevent the conditions that cause failures.

This experience-based understanding goes beyond reading a manual, making it a significant value. In the old days, people would be responsible for certain machines and would develop this understanding. Now people don't have the luxury of being responsible for just one machine. The need for greater efficiency has forced mechanics to work on a multitude of equipment. However, mechanics still need to have a detailed understanding of the equipment they work on.

The next block, **Innovative solutions**, is an example of how these blocks are values and not a process. Mine managers who disregard innovative ideas in favour of the laborious, costly, and time-consuming methods of the past risk falling behind their competitors. Innovation requires risk, and this block embraces taking risks in a controlled way.

Problems and breakdowns are inevitable, but your response doesn't have to be predictable. Try using more durable materials, replacing parts with updated electronics, or changing a system to reduce machine stress. Mining equipment has a lot of dynamic structures, and we wouldn't see the improved technology we have now if people never tried innovative approaches.

Innovation often means simplification. Another book that influenced me is *Simplicity* by Edward de Bono, who writes about the importance of knowing something in detail in order to simplify it. Maintenance leaders and experts must be able to simplify the process for people.

"IN ORDER TO MAKE

SOMETHING SIMPLE,

YOU HAVE TO KNOW

YOUR SUBJECT VERY

WELL INDEED."

EDWARD DE BONO

The value block advocating **Going the extra mile for others and expecting the same in return** deals with the relationship between your operations, engineering, and maintenance departments. We always find friction between operations and maintenance. An adage in the mining industry is that "maintainers fix the machines while the operators #$%* [destroy] them."

However, these two departments don't have to be adversarial. If you go out of your way to help another department, you can expect the same in return. I had this mindset ingrained in me when I was a planner. My maintenance manager, Steve Rae, set the tone, saying we would go out of our way to help the operations guys. We didn't let production walk all over us or disregard the established processes, but if there was something that needed attention that was in our power to fix, or if it was a little thing that would go a long way for them, we would pull out all stops to do it. If you throw up roadblocks—such as saying, "You're supposed to have the request in three days before lockout planning, so you have to get this sorted out yourself"—then you can expect the same opposition from them. However, if you say, "This is outside of our planning process, but we'll get it done for you," then make sure they are satisfied with the work, they will pay it back in spades.

The block about **Open and honest communication** is

a value I learned working at a mine site called Bengalla. The people at that mine didn't shy away from having hard discussions and stating the facts. The operation's commitment to two-way conversations allowed people to resolve problems. Another good resource in this area is the book called *Bullshift* by Andrew Horabin. This book offers solutions so that people can communicate more effectively and resolve issues that lead to tensions.

Open and honest communication prevents issues from becoming disruptive or emotional. Leaders need to present this as a value to their team and allow team members to become comfortable with the process. You don't want to hear about a communication problem for the first time when an employee tells you they are quitting. That's a failure. You want an environment where people are comfortable talking about any bad feelings they are having.

IMPROVING STEP BY STEP

The Maintenance Improvement Model includes several steps we must climb to reach our goal. You can't reach the top in one step. It requires a series of continual-improvement steps. The list on the right helps us see what we must adjust, change, and improve. These elements are specific to maintenance.

The first thing required for continuous improvement is

Benchmarking. How does the availability score for your site's truck fleet stack up to other mines? Is the reliability of your drills better this year than last year? Sometimes it is difficult to benchmark against other companies or mining sites; one company may calculate their performance differently than you do, and their mining operation or ore deposit may be much different than yours. In such cases, it's best to benchmark against yourself.

When benchmarking, think of your best-performing machine and ask yourself, "How can I make that machine even more dependable?" Similarly, if you benchmark against others and find you're the best, think about how you can do even better.

The next thing to consider is a **Breakdown analysis,** also known as root-cause analysis or a defect-elimination process. This is when you analyse why a machine broke down and how to prevent future failures. It needs to be a simple process, not complicated by heaps of written material and data processes with different steps and thresholds. Focus on improvement and avoid overanalysis that takes time away from implementing changes.

This connects to another point on the list, the **Top 10 Pareto analysis.** Vilfredo Pareto developed the 80-20 rule, which essentially says that 80 percent of the effects are caused by 20 percent of the causes. A Pareto chart helps prioritise the top ten problems. Ideally, you will want to chase, to the failure mode, every single breakdown in the top five. When you look at areas of a plant that cause difficulties, you will want to refer to Pareto charts to visualise what areas you should focus on. We'll look more into the practical process of this in part II.

The next point, **Maintenance delay analysis,** is something people don't do very well these days. People focus primarily on breakdown analysis but rarely examine delays in scheduled maintenance. Why were there delays in the service? Were the correct tools not delivered to the site? Did the mechanic fail to show up? Was the machine not set up properly? If you don't find what caused lost time, you can't improve the planning.

Most maintenance key performance indicators (KPIs) measure whether the work was completed and don't analyse delays unless something prevented the job from being completed. That's unfortunate, because analysing delays can lead to greater efficiency even if the job was eventually done within the scheduled window.

For example, some years ago at a plant in Indonesia, I asked the supervisors for their maintenance delay forms. They said they didn't have any forms because they did not have any delays.

"Absolutely we are having delays," I said. "We're not that good at planning and organising ourselves."

I went into the plant and talked with one of the electricians. He also insisted there were no maintenance delays. When I asked him what he was currently doing, he said he was waiting for a conveyor to be isolated. He had been there for two hours.

"Isn't that a delay?" I asked.

He said, "No, it's normal. It happens all the time."

I had to explain to everyone what a delay was. Delays should not be accepted as normal, and they can be avoided with improved planning and organisation. There's never going to be a perfect plan, but you can improve significantly by understanding what caused delays.

The next step, **Safety observations**, also known as field leadership, involves identifying areas where people disregard safety procedures. Nearly all companies have a version of this. Done correctly, safety observations have been fundamental in the mining industry's safety improvement record.

I did a safety observation during a manual handling task to change a starter motor. I noticed that people didn't have the right lifting equipment, especially for the tight space, and the starter motors were way heavier than documented safety standards of twenty to twenty-five kilograms. We identified procedural improvements and a different lifting device that would help us in the future to remove the starter motors safely using the overhead crane. More importantly, though, it made us realise that people were not aware of our current standards for these types of manual handling tasks.

Area inspection correlates with safety as well as efficiency. Ensuring your work area is in order is an essential step in an improvement plan. I would have daily inspections of our workshop around a simple set of checks:

- Are all things in their place?
- Are people following procedures?
- Are all the isolations in place?
- Are tools stored correctly?

These inspections create a controlled environment and complement any system such as Five-S. Without the daily discipline to ensure standards are maintained, order usually degrades.

Cost review and analysis reinforces the importance of a sound maintenance budget and reviewing performance against that budget. Depending on how actual costs are reported from the Enterprise Resource Planning (ERP) system, compared to how the budget was created, this can be easier said than done. However, it is essential to review costs against both the budget and benchmarks. The methods for doing this will be explored in later chapters.

Job observations, like safety observations, examine the quality of work and whether quality standards are followed. As with safety observations, this must be a self-auditing process. For example, if you are observing how a crew changes an engine, you can see whether people follow the engine change procedure and quality checks. Do they have all the parts bagged and tagged? Are critical bolt-torquing steps verified? With self-audits, you can self-correct as you go and prevent failures.

Maintenance-Driven Reliability Visit

RELIABLE EQUIPMENT AND RELIABLE BUSINESS RESULTS

"All Equipment Failures Are Preventable"

Area: _____ Date: _____

Reliability Visit Team: _____

Sample questions that you might ask during a reliability visit (there are many others):

Category A: Specifications

What are the specs? What are the torque settings for fasteners?
What are the dimensions? What are the tolerances?
Is there a plant standard? Are standards being adhered to?
Is there an ISO cleanliness standard for lubricating fluids?

Category B: Practices

Is best practice for this task being used?
Are fasteners tensioned correctly? In the correct sequence?
Are parts (elec + mech) cleaned before reassembly?
Are contamination control practices being applied?
Do we take oil samples from a valid sample point?

Category C: Tools & Equipment

Are the right tools for the job being used? Are they being used correctly?
No shifters!
Do the tradespeople have to improvise?

Category D: Procedures and Processes

Are job instructions adequate? Do they contain all the necessary information?
Is the job procedure being followed? What are the critical steps?
What can be done to improve the documentation?
Can the task be done in a better way?
Was the equipment prepared on time and in the necessary condition?
Are people aware of the fundamental processes (equipment strategy
    development, work management, reliability improvement)?

Category E: Strategy

Why, why, why, why, why.
Why are we doing this task? Why are we doing this task this way?
What things do we find when doing this task? What can we do to prevent them?
Can we avoid doing this task?
Are we changing components before their targeted life?

Cat (e.g., A)	Discussions Held/ Observations Made/ Recognition Given/Action Taken	Potential Business Impact	Follow-Up by Whom/When

Classification (Class)	A	B	C	D	E	F	Total
1 – Good Practice							
2 – Good Condition							
3 – Poor Practice							
4 – Poor Condition							

Number of Corrective Actions Identified

Total Number Identified	Number Recorded in SAP?	Number Recorded in Another System?

"All Equipment Failures Are Preventable"

Another crucial aspect of my model is the wedge on the left. The wedge represents practices that prevent the maintenance system from falling back down the stairs.

The first practice in the wedge is **Document Control**. Maintenance departments rely on written procedures and standards, and if those documents aren't managed properly, people can work off outdated information with missing details. Good, simple document control processes enabled us to improve our consistency significantly during the nineties when ISO 9001 was first being adopted in the mining industry.

Safe Work Procedures (SWPs), also called safe operating procedures, must be practically implemented and understood by the people who execute the work. Allowing workers to develop these procedures fosters understanding and commitment.

Maintenance Strategies means different things to different companies. People call them "asset management plans," "strategies," "tactics," or a combination of them all. Terminology has been adopted across different companies. The "Asset Management Plan" is a detailed, equipment-specific document on how to maintain a machine across its life cycle.

The most critical element is the scheduled downtime strategy. How often and how long do you shut down the machine for scheduled maintenance? What do you do while it is down for scheduled maintenance? What standards do you expect? What should be greased? What should be cleaned? What are things you should adjust through scheduled maintenance to prevent the onset of failures? There's a lot that must be included.

After that, there's **Training and Development**. It's important to consider how you will train people in all of these documents or make certain that people understand them. Whenever I examine how an organisation is doing at this, I ask for the training matrix. Managers typically have a training matrix that focuses on the skills employees need to start working on a site—how to isolate, work at heights, and all the critical core safety skills—but not much related to detailed, equipment-based skills. Mechanics might be skilled with a piece of equipment or fault-finding on a type of machine or plant. But detailed, technical skills need to be rigorously included in a well-rounded training matrix so that every crew has someone with knowledge of a particular plant the team maintains. A training matrix helps supervisors ensure they have the necessary skills on their team, but supervisors must recognise it is their responsibility to do this.

Rituals also help keep the ball from falling. This means having daily routines, such as workshop inspections, daily discussions, and reflection on learning, cleaning up, and setting up maintenance tools and staging areas for vehicles and parts. Every item in a workshop must be labelled. Maintenance departments require this high level of organisation, and the rituals help maintain that organisation and create the culture. Staying conscious of everything that needs to be documented and discussed day after day keeps you from falling backward.

PUTTING IT ALL TOGETHER

If you're wondering if your maintenance team has the right attitude about the work and the correct behaviours to help make your team successful, take a step back and see how your team stacks up against the values and strategies in this model.

Does your team have an innate desire to improve? Do team members go the extra mile? Do you all practice open and honest communication? Are you adamant that scheduled maintenance has the highest priority, and are you willing to let some breakdowns wait for resources in order to realise that value?

If you can say yes to these and the other challenges posed by the Maintenance Improvement Model, then you are well on your way to having a successful team culture.

However, chances are you are falling short in some of these areas. Use this model to document the values that make sense to you and your team.

Whatever your situation, keep that first step in mind: the desire to improve and the belief that you can make a difference. If you and your team have that, it's more likely the other bricks will fall into place and you'll have greater success getting that ball up the stairs.

Values Foundation/Bricks

Insert those that mean something to you and your team

Desire and determination to improve

THE WORK MANAGEMENT MODEL

This model is a simplified version of the common Pipeline Model. It illustrates the flow of work from the strategy that drives our activities to the continuous improvement loop.

This process is generally accepted as the maintenance process no matter how people prefer to see it written. In fact, people always have different views on how the process should look, but in the end, it is the same process.

Work Management Model

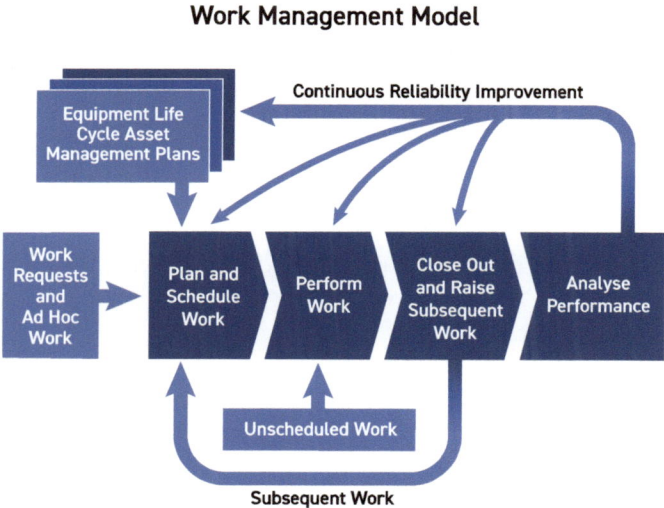

Most organisations have some form of Pipeline Model that shows:

- Planned work
- Scheduled work
- Work performed
- Work completed
- Work closed out
- History
- Subsequent work captured
- Data analysed for performance improvements

Many of these models have a section called "identify work," but I've removed it from my model because identifying work is part of performing work. As you perform a task or an inspection, you should identify other work that can be done during the next scheduled downtime event.

My model includes "work requests and ad hoc work" in the first chevron of the pipeline. These are work requests from other people, original equipment manufacturers (OEMs), statutory bodies, or operators. This work is for a specific purpose and is not driven by a strategy or asset management plan.

Work originating from asset management plans or equipment life cycle plans should be set up when you purchase and initiate the machine on site. These are jobs you expect to do throughout the machine's life, such as changing components according to time or condition. The other way that work enters the planning and scheduling process is from the completed inspections.

Once jobs are planned and scheduled, teams perform the work while accommodating unscheduled work or breakdowns. When the work is completed, there must be a disciplined close-out to ensure people know what is going on in their business (i.e., understand the details of the job status and what was done). I once took my team to a high-performing site so that we could learn how the site managed its operations. The supervisor at the site showed how he entered job history into the computerised maintenance management system (CMMS). It was a lot of extra effort, he said, but it paid off because it helped everyone understand what was going on. That process was the difference between their site and ours; we did not know what was going on, because we were not capturing the information with discipline. We adopted that process and it immediately improved our operation.

Closing out work and capturing the history or information about what was done are critical when there are many crews and lots of people involved in the teams. Without common understanding of these standards, many issues will be missed and become breakdowns.

The next stage is to analyse the performance. The information, including any maintenance-delay analysis, downtime information, and work-done information, facilitates improvement. In chapter 4, we will discuss this process and how to make it simple and effective.

IMPLEMENTING THIS MODEL

This model is the highest level of the work-management process. Each box has a detailed workflow process associated with it. The work requests and ad hoc work, for example, could drop down to another work process that includes "How the operator puts in the work request, which then goes to the supervisor. If the supervisor approves it, it goes into the system. If not, it goes back to the operator."

These workflow processes have been repeatedly documented in the mining industry since around 2000, so they are quite detailed. They are useful processes to refer to so people on a team—particularly at the management level—can utilise the system consistently. However, it's counterproductive when companies try to force every site to use the same precise processes at a detailed level. It's better to give onsite tradespeople and leaders some flexibility to do what they know is best, even if they don't follow the process to the letter. What's important is that they achieve the agreed outcomes, which can be as simple as ensuring everyone knows what is going on!

Even when sites document workflow processes to the nth degree, some variations still exist across different sites and different people. However, these variations shouldn't affect equipment performance.

Outputs at each stage of the workflow process are not negotiable. For example, if I look at the work-request process and can see that the operators are satisfied, I know things are getting done when they are reported. If, on scheduled work, the parts and tools are delivered to the job on time and the workers know what they need to do and when to do it, then I know the planning and scheduling process is working. If I don't see any defects captured in the CMMS—but I can walk out to the plant and find leaks, loose parts, dirt buildup in electrical cabinets, or gears that aren't lubricated properly—then I know there is a problem with execution or the close-out and capture of subsequent work processes.

The main point here is that if you make process manuals complicated, people can't or won't read them or understand them. The Work Management Model is simple and more than adequate to get control of the basics and deliver the results. Focusing on the outputs at this higher level is simpler and reduces the complication of administering the work-management processes.

IF THE TEAM HAS A CULTURAL DESIRE AND DETERMINATION TO IMPROVE, THEY WILL SORT OUT THE DETAILS, CREATE THIS LEARNING CULTURE, AND WATCH THE EQUIPMENT PERFORMANCE FOLLOW!

CHAPTER 3

Focusing on the Major Pain Points

This chapter explains what compelled me to discuss these models in relation to the problems faced by those working in mining maintenance. I will also review five major pain points with which you will be familiar. I will go over each of these points in their own chapter in part II.

A COMMON UNDERSTANDING

When I talked about these models in the previous chapter, my focus was on using them to align understanding to overcome common problems. This approach is similar to the way a successful sporting team works. The coach and team members develop and commit to a game plan. The game plan is a model showing each player what to do to achieve the goals and win the game. Maintenance managers are the coaches of their teams, and the managers need to communicate concepts so team members have a shared understanding of how to overcome a common problem.

All maintenance managers face similar problems. Managers are continually reacting to problems, even on their days off. It wasn't always that way. I remember a time when maintenance was more organised, the schedule less chaotic, and the manager's role much less intense than it is today. Today, maintenance managers are always worried about the phone ringing with another disaster.

Supervisors are in a similar reactive state. They are dealing with surprises or people issues all day, every day. They have to stay late to finish up the day's work, and many come in early to work in peace without the distraction of other people.

CEOs and COOs, meanwhile, are focused on cost, production targets, and safety. They have to get results. The general manager is responsible for implementing all the strategies and actions the CEO and COO use to hit targets. The CEO and COO deliver these targets, but the general manager must meet them.

Unfortunately, the general manager and the maintenance

manager often have a broken relationship. That's because their key performance indicators (KPIs) are different and are often in conflict. The maintenance manager is focused on making sure the equipment is reliable, the general manager is focused on profits, and the general manager often doesn't see how equipment reliability is going to help him reach his profit goals. This creates a disconnect.

THE PAIN POINTS

The most significant problems can be categorised into five areas:

- Excessive Unscheduled Downtime—This is the result of too many breakdowns, causing maintenance teams to be reactive.
- Excessive Scheduled Downtime—We must have scheduled downtime, but if not managed efficiently, it will cause excessive production losses.
- Cost Concerns—These are problems resulting from not meeting a budget, not controlling costs, or needing to reduce expenses to keep the business viable because of low commodity prices.
- People Problems—High turnover of employees can be the starting point for many other problems.
- Project Transitions to Operations—These problems surface when a mine or engineering or construction project transitions to the operations phase.

You may find that there is some overlap between these pain points. The unscheduled downtime will also cause cost concerns. Similarly, the solutions will overlap with the values and business processes we talked about in the models. For example, the solutions to excessive scheduled downtime may include the lack of a delay analysis, or people don't accept responsibility for optimising scheduled downtime. In that sense, the models provide a context for discussing these pain points.

I don't necessarily want you to read the second part of this book sequentially from chapter 4 to chapter 8. Instead, go to the section of the book that addresses your current headache. If your machines are reliable but costs are a problem, jump straight to the chapter on costs.

The key point is that **this is a people business**. People need to work together with open and honest communication and a shared approach to achieving the goals.

There is no technical process that will fix these problems, no matter how many times we document it. Computerised tools aren't going to energise an unmotivated worker. One of the things I'm proud of with my business, Bluefield, and an essential reason for its success, is our consistency. Most of the people who started with us eight years ago are still with us, and we always emphasise precise communication and taking responsibility. We have arguments and hard discussions, but that is necessary to achieve mutual accountability. We can fix our problems because we look at ourselves first. When something goes wrong on the job, rather than point fingers at those responsible, we always ask ourselves, "What can we do better in the future to prevent this problem from happening again?" Our business is steadily growing because of this collegial approach.

PART II

Common Problems and Practical Solutions

TRAITS OF A SUCCESSFUL TEAM

When I review a site where the equipment is not performing well, I often find great individuals but flawed teams. Here, I'll list the three elements all teams must have, and I'll put a tick (✓) next to the elements that I see often, and an X (✗) next to elements that are often missing.

- **People with complementary skills and experience. ✓**
  I always find this element. Sites have mechanical experts, electrical experts, skilled planners, engineers, and managers.

- **People committed to common performance goals.** ✓
  I always find this element. There may be some differences on some metrics, but everyone wants to achieve good reliability and business performance.

- **People who are committed to an approach for which they hold themselves mutually accountable.** ✗
  This is a common problem. This occurs when people don't talk about problems or buy into a philosophy that stresses taking responsibility. Team members don't feel a common sense of duty.

These three elements appear in all the solutions I present in the following chapters. Each chapter examines an overarching problem and what managers can do to address them practically and sustainably. The common thread that runs through all these solutions is the idea that teams must be aligned with each member's job and responsibility. This alignment can't come from a manual; it must come from frequent discussions, which develop a successful mindset. Solving problems this way is much easier and more effective than developing a bureaucratic solution.

TECHNOLOGY ALONE DOES NOT HELP

Over many years, I have seen people try to solve these problems using technology. Unfortunately, technology is rarely a solution to the problems described below. Technology is a great benefit and must continue to advance, but if the basics described below are not done well, even the advanced technology will fail.

There are many examples of new technology in our industry that have failed to deliver the expected benefits.

In one business, there was a project to automate the warranty management process. Warranty claims were not being processed, and the business was losing money as a result. The site decided to automate the warranty management process in SAP. It cost $2 million to implement the automation process. Site people only had to push a button to initiate the process.

After two years, the CFO asked me why the site was turning off the warranty automation software tool. I investigated and found they were turning it off because it was not adding any value. No one had initiated any warranty claims in the system.

This was a clear example of why a system will not work if the people are not engaged. It is even more important now as we head into the world of using artificial intelligence and predictive analytics. These technologies are great, but there must be a culture of quality work execution to support these technologies.

CHAPTER 4

Excessive Unscheduled and Reactive Downtime

Addressing breakdowns is the first step to improving maintenance. Many managers think planning and scheduling is the first step to reducing unscheduled downtime, but planning and scheduling only helps once the basics are in place. If these fundamentals are not developed first, planned and scheduled work will always be interrupted by breakdowns. This chapter will focus on the areas that will have the most significant impact on reducing unscheduled downtime.

WHAT IS THE CAUSE?

There are many explanations for why mines have excessive unscheduled and reactive downtime. I will examine eleven of the most common reasons. The overarching reason for breakdowns is a failure to take ownership and care of the assets. It's what I call a lack of "execution quality." Unexpected breakdowns result from poor-quality maintenance execution, incorrect operation of the equipment, or a lack of overall care. When I talk about maintenance execution, I refer to the physical execution of the work, including preventive tasks, condition inspections, and the process of work-order close-out and raising of subsequent work.

Here, in no particular order, are examples of the primary problems that lead to poor execution quality, and what managers can do to address them.

DEFECTS DON'T GET INTO THE SYSTEM

Maintenance managers can avoid mishaps that cause reactive downtime if their tradies identify problems during scheduled services and record those defects in the computerised maintenance management system (CMMS). If the people doing inspections find faults, they can't just note them on service inspection sheets; they must put the issue in the CMMS so planners can schedule repairs. Otherwise, no one knows about the defect until there's a breakdown or at the next scheduled inspection, when it is found to be out of specification and needs to be replaced immediately. This puts people in a reactive mode because they do not have the resources ready to correct the conditions.

Here are two real-life examples of this problem. These two examples show service sheets listing serious defects. Mechanics reported these problems, but there were no subsequent notifications or work orders raised in their maintenance systems. No one addressed the problem, which could lead to breakdowns or a fire on the equipment.

Defect Example

COMMENTS

Service leg mounting pin came out
RH stick cylinder stauff clamp missing off rod end hose
LH engine fuel return hose cracked
LH engine radiator door has 1 handle U/S

Check Trunion Bush			
Check Trunion Grease Supply	✓	✓	✓
Check Fuel Lines Wiring harness starting to rub on HP fuel line LH engine	✓	✓	✓

The first example is a failing fuel hose. Someone inspected the engine on a mining truck and noted that the fuel hose was wearing out. A cracked fuel hose could leak and spray fuel onto the hot parts of the engine and cause a fire. This worn-out fuel hose wasn't raised as a defect in the system nor repaired immediately during the service.

There could be many reasons for this. Perhaps the inspector forgot the issue after making many similar reports, or the inspector assumed someone else would make the subsequent notification. Whatever the reason, no one recorded the problem in the system, and no one took responsibility for ensuring corrective maintenance was scheduled.

I've seen this issue on many occasions. An inspector notes a small leak on one service sheet, then a more worrisome leak appears on the next service sheet, and then the equipment breaks down a few days later. The malfunction leads to damage, higher costs, lost time, and the potential for injury.

The other example is from the same site. An electrical wiring harness was rubbing on a high-pressure fuel hose. An inspector noted this problem but, like the first example, never put it in the system.

Due to their potential consequences, these examples

of unactioned defects are among the worst I have come across. However, this kind of event is common, and many mines have no clear system to prevent them. In a small organisation where a maintenance person makes a mental note of a problem and ensures it's fixed before it worsens, there may be no need for a system-based solution. However, a company with large fleets of equipment can face serious consequences if these issues persist. Large companies need reliable systems that everyone uses so simple repairs don't lead to debilitating breakdowns.

The first step to solve this problem is for everyone to agree on a method to overcome the problem. This method might be as simple as writing a notification number next to every defect that is identified so that someone else can verify that it gets done every time the service is performed. The specific solution is not as important as the fact that everyone agrees to do it the same way so that we all know what is going on. In Bluefield, we call these small agreements within the teams *working agreements.*

Developing these working agreements requires getting all of the crews together from different rosters, particularly the supervisor from each roster and the people who form part of the natural work team. This collaboration can also include the planners and engineers for that team. The leader (superintendent) must be part of the discussion.

Everyone must feel that they have had an opportunity to provide input. The decision can't be made by supervisors or superintendents and delivered to crews. Crew members need to own the solution. If you just tell them what to do without enabling them to decide on the solution to the problems, it takes the ownership away from the people who have to implement the solution. Removing ownership this way makes the leaders' job more difficult because now they must hold crew members accountable for something the crew members may not see as necessary. However, if there is ownership across all shifts, things will change—even if the working agreement is just scribbled on a piece of paper.

The goal is to ensure that all defects are consistently and verifiably recorded on service sheets and acted on. Some defects do not require subsequent action, such as when there already is a job for it in the system, but the team needs to be conscious of this.

A tablet-based inspection system is a more technology-based solution. This approach is just emerging, and it shows promise. It can automate the process of raising subsequent actions, but the technology is still being developed and not many companies have yet adopted it.

This is when defects are overlooked or ignored. An inspector following an inspection sheet telling him to look for rubbing hoses, leaks, or loose components reports that everything is good when it isn't. The example below shows this, and it is not uncommon to find these conditions.

Inspection Sheet & Rear of Truck

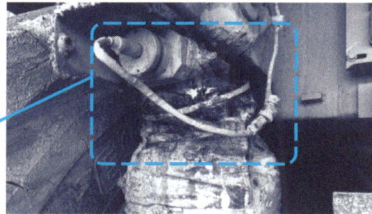

Required Inspection

Condition after the inspection with no subsequent defects raised

There could be many reasons for this:

- The inspector didn't notice a defect or is unaware that the condition is substandard and might cause a breakdown.
- The technician has a bad attitude and doesn't care. This is very rare but can occur on sites with morale problems.

- The technician assumes the poor condition of the equipment is acceptable. People generally adopt the standards they find when they arrive, so if those standards are low, that's what the inspector will allow. It requires the team to make it clear what is acceptable.

Some inspectors overlook problems because they don't recognise these defects are causing breakdowns. Sometimes the standards are written on the service sheets and inspectors still don't identify the defect. But most often, unidentified defects reflect the low standards adopted by maintenance crews at a particular site.

Again, the solution to this problem is to get all people to agree on what standards are acceptable. Maintenance crews need to discuss the quality, standards, and conditions of the equipment every day. Some of the sites that have achieved the most rapid improvements in equipment availability (from 84 percent to 90 percent in less than twelve months) implemented a process where they shared photos of bad practices and good practices at their shift-start meetings. This ritual changed the culture because the mechanics doing the work wanted to be recognised for good work and did not want to let their team down.

Most mining sites have morning meetings where the main topic is safety. The same conversation can also focus on quality and standards.

Another way to improve standards is to set up a failed-parts inspection area. People can examine the parts and service sheets and discuss why a piece of equipment broke down. Everyone learns from the experience.

Some companies have a safety cross as part of their lean board, and these operations often set up a maintenance execution quality cross too. The key here is for teams to figure out their own way to raise standards; they are more likely to take responsibility if the solution comes from them and not from the top.

WORK GETS DONE TO POOR QUALITY STANDARDS

Although the previous problem is also related to poor quality standards during preventive maintenance (PM) inspections, this problem refers to the technical crafts-manship of corrective, preventive, or proactive work. Too often, someone replacing a brake assembly, hydraulic hose, engine, or some other component doesn't do an adequate job. There is no good excuse for substandard work. The OEM provides mechanics with a procedure for doing the job correctly, so there should always be specific quality information to ensure bolts are torqued correctly and that there is proper fitment and setup.

Welds require particularly high-quality standards. Welds are prone to recracking if correct pre- and post-heating standards are not employed and the weld surface is not finished properly. However, many mine welders rarely consider the required quality standards.

Years ago, we welded keepers on dragline bucket teeth, chains, and pins. These are all bolted today, but they used to be welded and they frequently would fall out.

There wasn't complicated root-cause analysis back then. We just engaged the mechanical engineer, who worked with the welder and figured out that the keepers weren't preheated before being welded. We made a change to the acceptable standard, and when the welders started preheating in this way, the unscheduled downtime plummeted—all because we used correct welding procedures on a simple welded keeper.

These examples illustrate the value of having a culture of high quality.

Maintenance team members must also agree to follow the written procedures and documents, whether it is the remove-and-install (R&I) documentation, a quality assurance/quality control-type document, or a PM inspection checklist. In addition, maintainers must alert leadership when they think documents are wrong. Ticking boxes when the work is not done adequately is of no value. Team members must talk about instances when work is poorly executed. Everyone must care about quality. We do it once, and we do it right.

Temporary repairs must be done at times and are okay as long as you raise a subsequent action in the CMMS to ensure a permanent repair is made. Temporary repairs that are not raised and corrected permanently lead to future breakdowns.

We have helped many sites change the execution standards or change the culture that accepts low standards. There is not one way to do this or a magic bullet. It is more important that teams have a desire to improve and face reality; then the teams will find a solution. Some of the mechanisms that our clients have used to enable this cultural shift include using a quality cross to track execution standards and implementing a daily quality discussion into the shift-start meeting. In these meetings, it is important to recognise good quality execution standards and highlight poor quality examples so that the standards are continually raised. This process is very similar to the safety improvement journey that our industry has been on for twenty years.

In addition to raising the standards, this daily discussion allows tradies to talk about technical issues and improve their technical knowledge.

Below is an example of a quality cross and a quality standards photo board.

Maintenance Execution Quality
WORKING AGREEMENT DAILY UPDATE

Team: Mechanical Maintenance July 2018

		1	2	3		
		4	5	6		
7	8	9	10	11	12	13
14	15	16	17	18	19	20
21	22	23	24	25	26	27
		28	29	30		
			31			

GREEN No failures due to maintenance execution or work management discipline and improvement implemented.

ORANGE No failures due to maintenance execution quality or lack of work management discipline.

RED Failures due to maintenance execution quality or lack of work management discipline.

Monthly Top 5:

_____
_____
_____
_____
_____

Quality Photo Board

Quality Share – EX02 fan hoses rubbed on guard

Quality Share – EX21 Alternator belt broken shortly after install

Quality Share – EX31 Grease Faults

Quality Share – EX07 Dog bone grease lines not secured

Quality Share – Not setting up for success (generator)

Quality Share – Lack of information in work orders

TOLERABLE DEFECTS ARE NOT MONITORED AND MANAGED

Often you find a defect—such as a small crack that seems to be getting worse—that doesn't need to be fixed right away. The machine can continue operating, but you must keep an eye on these defects and manage the problem until the next scheduled service. When that comes, you can address it with adequate time and care. If the problem worsens, you can move up the date of the next scheduled maintenance.

Recently, we had a coupling that was making a loud noise and overheating. Workers filled it with grease, which cooled it down and stopped the racket. I told them to keep operating until the next shutdown in ten days but to examine it frequently.

They treated it as a tolerable defect. They inspected it every couple of hours, and as their confidence grew in the temporary repair, they decreased the inspection frequency. If it started rattling or overheating again, they shut it down, added more grease, and kept going. All crews were aware of the problem and kept an eye on it, and the coupling made it to the next scheduled outage.

It's okay to manage these flaws until you have the people and parts to fix the defects during a scheduled service. Every site has these defects, but they often are not managed well, such as when the roster changes and the departing team forgets to warn the arriving team about the defect and how to manage it. A breakdown inevitably occurs.

The solution is to have an area on the shift-start agenda—a whiteboard or sheet handed out to everyone—that lists all the tolerable defects being monitored. Everyone must know the condition of the machines and be aware of potential defects that could cause a breakdown if not managed. The list should define who will check on the defect and how often. Crews should time the inspection to coincide with another downtime event, such as refuelling. Good communication can prevent a costly breakdown.

Tolerable Defect Management

Work Allocation/Shift-Start Board

SCHEDULED WORK

Location	Unit	Description	Who	Light Vehicle Allocated	Scheduled In	Scheduled Out
R2	209	Service + Defects + Adaptors	Fred, Mike, Bluey	`12-4 `12-6	6:00 a.m.	6:00 p.m.
R8	109	Mech + Elect Inspection, Injector test	Smithy, Joe, Pete	`12-8 `12-9	6:00 a.m.	6:00 p.m.
Shutdown Pad	231	Scheduled shutdown	Contrator		21-Mar	26-Mar

BREAKDOWNS

Location	Unit	Description	Who	Light Vehicle Allocated	Estimated Out
R3	108	Replace fuel injectors	Dave, John	`13-11	3:00 p.m.
R4	111	Slow carousel movement	Sam	`13-6	TBA

TOLERABLE DEFECT MANAGEMENT

232	Monitor book crack. Inspect daily at refuel.
234	Monitor main hyd pump leak. Inspect daily at refuel.

JOBS IN THE SYSTEM GET LOST AND AREN'T COMPLETED

This problem occurs when someone raises a defect in the system, but no one schedules the work, and a breakdown results. This often happens when the people looking at the backlog/forward-log overlook the defect. Sometimes they filter the open work orders in the system and do not look at all of them. Some fall outside of their filter criteria and become lost in the system.

One time I saw an "out of service" tag on a dozer ladder. The dozer was in for scheduled maintenance, but the broken ladder wasn't part of that repair plan. I learned that the ladder had been out of service for six months, but planners were searching for work orders only from the last three months. They didn't see it. Not only could this have resulted in an unscheduled downtime event, but it could also cause operations to lose confidence in the maintenance team.

At some mines, jobs get lost in the noise. Some jobs are completed but not closed out. Some people don't know what's what.

To test for this at a new site, I pick a random machine or part of the plant and organise a meeting with the planner, supervisor, superintendent, and tradespeople to review every single job that's in the system for that particular machine or plant area.

I don't expect one person to know everything that must be done to each machine. However, for each job in the system, at least one person from this group must know the details. As a team, they should be aware of everything.

It is not uncommon to ask about the status of each job and hear lots of maybes or nothing at all. Silence. No one knows. There are often work orders that have been raised, and though no one can say whether that work was completed, it seems likely that it was. Looking at all of the work orders in the system like this can be, for the assembled group, a realisation that their process to "close out and raise subsequent work" is out of control.

Once again, everyone must be on the same page to solve this problem. The maintenance supervisors, superintendent, planner, and operations team must agree on a plan and priorities for the next scheduled period and be sure that all jobs are reviewed. The maintenance and operations team must go through the backlog or forward-log every week.

This approach may sound like telling someone that the solution to not suffocating is breathing. Unfortunately, these basic problems repeatedly occur. Meetings and detailed procedures won't help; people won't be motivated to solve these problems unless they take ownership of the process and the outcomes.

SERVICE SHEETS DON'T SET PEOPLE UP FOR WORKING IN A SCHEDULED MANNER

Some truck service sheets say brake pads must be replaced when they're under twenty millimetres thick. I have seen instances where brake pads were recorded on the service sheet as being 20.1 millimetres thick, and no one did anything because it was above the limit. At the next service, however, the machine came in with brake pads that were under specification and people had to scramble to find labour and parts. It became a reactive job for them and created more unscheduled downtime.

A solution is to set limits for scheduling work and limits for immediate action. Keep twenty millimetres as the limit for an immediate break-pad change but establish twenty-one millimetres as the threshold for when brake pads must be scheduled for the next service. The brake pads will be below twenty millimetres by then, and if the work on them is scheduled, you avoid unscheduled downtime.

Another solution is tablet-based service sheets programmed to automatically raise the subsequent work orders when the limits are reached. These are emerging, but there is a large task still to be done in order to get the data into these systems.

Acceptable Limits

Cylinder

12. Estimate and record the pin-to-bush movement and note the locations:
Location 1: [Stick L/H] mm 3-4 mm

Movement < 3mm = Monitor
3mm < Movement < 5mm = Schedule replacement
5mm < Movement = Replace promptly

Location 2: [Stick L/H] mm 3-4mm
Comments: Both ends have movement

MAINTAINERS WANT PERFECT SERVICE SHEETS

Although you want service sheets that set up mechanics to work in a scheduled fashion, you can't have a perfect service sheet. You still need people on the floor to use critical thinking and consider what's best to do in a situation. If brake pads get down to 20.1 millimetres and inspectors know the limit is twenty millimetres, they should be able to deduce that it should be scheduled now because the pads will be below the limit by the next service.

The airline industry has highly detailed service sheets. For example, when checking leaks, inspectors are asked to count the number of drips per minute and compare the result to acceptable limits. Airlines also require general-area inspections that look for other defects not specifically documented on the service sheet. These general-area examinations work because the inspectors take responsibility and have a sense of ownership of the equipment. When you don't have that sense of ownership and inspectors are only checking off the items on their service sheet, critical thinking goes out the window and problems are overlooked.

Maintenance managers should always try to improve their service sheets, but they also must cultivate an environment where people take care and look for other problems. Service sheets are developed with the assumption that the equipment will work in certain conditions and will be operated in certain ways. However, inspectors may find that the equipment is not being used correctly or is operated under the wrong conditions, and it is their responsibility to look for the unexpected consequences.

For example, if you have a car with a slide-out drink holder, the expectation is that people will slide that out, put their drink in it, and push it back in when they're done. The maintenance program for the car might not have the drink holder checked for ten years or more if it's used in this way. However, if someone uses the drink holder as a footrest, it will show signs of wear much sooner, so whoever is maintaining that car needs to look for that.

An example from the mining industry is the loading arms of a wheel loader. These arms are designed so a loader can approach loose material with the front of the bucket. The arms are designed to last the full life of the machine, so the service sheet may call for only an annual inspection of the arms.

However, some operations use a wheel loader on material that hasn't been adequately blasted. When the operator approaches the material at an angle with the corner tip of the bucket, the action places unanticipated stress on the arms. That kind of stress causes the arms to crack. Mechanics must use critical thinking and examine the arms more frequently at sites using the wheel loader in unexpected ways.

MAINTAINERS ARE NOT TAUGHT SUFFICIENT TECHNICAL SKILLS FOR MACHINES

When you have a problem and it's not immediately apparent what it is, you do what's called fault-finding. You run tests and do some checks. If you're a mechanical guy with a hydraulic system problem, you do some pressure tests. If you're an electrical guy, you check currents, voltages, and control signals.

However, these workers often are not taught how a machine operates so that they can accurately diagnose problems.

I recently saw a guy find a leak in a gearbox. He said it required a new seal. They ordered the seal and installed it at the next service. However, that seal wasn't the problem. The guy who did the check didn't have enough technical knowledge of that gearbox to correctly diagnose the source of the leak. This kind of misdiagnosis can continue until the machine breaks down and needs to be repaired in a reactive, unscheduled fashion.

If the maintainers had better technical knowledge of the machine, the problem would have been correctly diagnosed and fixed the first time. Likewise, if the seal wasn't installed correctly because the mechanic didn't have the right technical knowledge, it could fail prematurely.

To correct this, supervisors must take responsibility and make sure members of their teams are balanced and have varied skills—people who can fault-find a hydraulic circuit, people with experience on specific electrical systems, and people with specialised skills for certain types of equipment. If you have a team of ten people, the supervisors must ensure there are adequate skills across this team to cover all technical needs, even when some team members are out sick or on leave. If the supervisors continually find themselves without the specific skills required for a job, they must accept the responsibility and take steps to correct the situation.

The training and HR departments can help develop workers and provide training, but the supervisor is responsible for the capabilities of team members. They must ensure their team has the skills required to do the right work on the equipment with the right quality. It is also the supervisors' responsibility to put people in situations where they can get hands-on experience, and skills transfer from other specialists in the team.

THE TEAM TALKS ABOUT PROBLEMS BUT DOES NOT TRACK AND MEASURE IMPROVEMENT ACTIONS

When a team discusses difficulties and breakdowns but doesn't track the actions required to avoid them in the future, jobs don't get finished. The team doesn't lock down who is responsible for the corrective action and doesn't document the action to ensure it's completed. Sometimes the actions are captured well, but no one monitors them.

We've all been in meetings where people agree to take action on a maintenance issue, but as soon as they walk out, they are swamped with other requests and forget. No one feels satisfied working in this environment; maintainers genuinely want to complete the preventive work and avoid repeat failures, but urgent tasks get in the way and prevent that from happening.

Teams also need to determine what kind of training team members need and set aside time to ensure the training occurs. Teams must have an action-tracking system, which is simple to do with the ERP or CMMS. Those actions must be reviewed on a daily or weekly basis to ensure improvement occurs quickly enough.

For example, a hydraulic system breakdown once lasted a couple of days as mechanics struggled to fault-find effectively. When the problem was finally discovered, the fix took only an hour.

The action for this situation would be for someone on the team to research options for training on fault-finding this hydraulic system and bring those options to the team so it can decide whom to train and when the training will occur. This action plan can't be put into a work order, but it must be tracked to ensure it is completed. The training has to be scheduled, put on people's calendars, and not postponed. This is not an urgent task, but if this training is pre-empted by the little emergencies that pop up during the day, the team will have the same problem again.

EVERYONE THINKS THE PROBLEM WITH RELIABILITY IS DUE TO THE RELIABILITY ENGINEERS

When I started in the industry thirty-three years ago, there were no reliability engineers. It remained that way for the first eight years of my career. The industry gradually introduced reliability engineers and charged them with fixing the little issues causing unscheduled breakdowns.

Unfortunately, many people now think it's the reliability engineer's job to fix all these issues with no involvement from the people executing the maintenance. In reality, these problems are often caused by factors—such as poor workmanship—that are outside the reliability engineer's control. Reliability engineers cannot solve problems without working closely with the people who respond to breakdowns. Both have responsibility for ensuring the reliability of equipment.

They must keep the parts for inspection or discuss the conditions found by the technician who went to the breakdown.

To correct this problem, we must recognise the joint responsibility for reliability between execution and the reliability engineering team.

This graphic analyses a breakdown event. The reliability engineering responsibilities are in light blue and maintenance execution responsibilities are in dark blue. The maintenance execution team is responsible for equipment reliability, and reliability engineers support the execution team. The process works when teams understand their responsibilities and never abdicate that responsibility to another team.

Improvement Verification Process

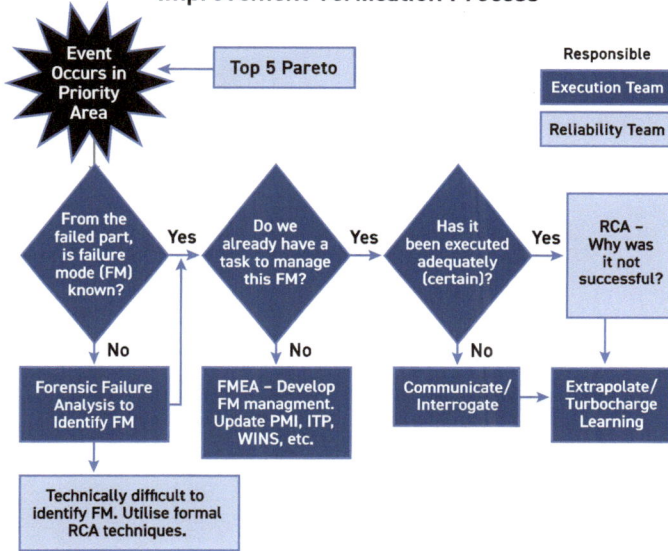

This approach improves the reliability of equipment. However, there are other reasons why reliability engineers are not effective. We wrote about these problems in our blog, which can be found at the following URL: https://bluefield.com.au/2018/03/14/the-top-9-mistakes-companies-make-with-reliability-engineers/.

OPERATIONS DRIVE REACTIVE MAINTENANCE

Maintenance needs a regular routine. If the general manager and operation manager suddenly change priorities, and the maintenance crew is forced to change its schedule, all of the effort that went into planning and scheduling is wasted. Maintainers become reactive, their proactive work is delayed, and the breakdowns caused by delayed maintenance and troubleshooting mount.

For example, say one of a site's four shovels is scheduled to go down for planned maintenance. However, when the scheduled shutdown time arrives, one of the other three shovels breaks down. Operations—not wanting more than one shovel to be shut down at a time—orders the shovel scheduled for maintenance back to work until the other shovel is fixed.

This approach hurts your operation in the long run. Scheduled maintenance requires a lot of planning. Sometimes it is necessary to hire off-site contractors who specialise in specific areas of the machine. You have to organise ahead of time when they will come in, and if you cancel at the last minute, you may not be able to reschedule these specialists for quite a while. So the work ends up not getting done at all, or done poorly, which starts the reactive cycle.

I have found that general managers typically get on board when they can see the ripple effect of these types of last-minute decisions. They just need to be shown how it impacts the ability to perform good quality scheduled maintenance.

When I was a supervisor, I came to work one day and

found a shovel down. There was already a shovel scheduled for maintenance, and the operations manager asked me, "Look, I know we're supposed to service that shovel today, but we need to get the coal out so we can meet our coal production targets for the month. Can we cancel that service today?"

I said, "I'm only a supervisor, so I can't overrule what you say, but I know that if we don't do this service today, we won't get to all these things that need to be fixed while we have the labour and parts. The next time, this machine is going to break down, and we'll go down this spiral, which will impact the production targets every month."

The manager understood. He agreed to continue with the maintenance as scheduled.

I have a short presentation I show executives on the need for scheduled service. I show how the reactive loop starts when we delay scheduled maintenance. The only way to prevent that is to remain committed to scheduled maintenance.

CHAPTER 5

Excessive Scheduled Downtime

Scheduled downtime is a good thing. Scheduled downtime, executed to adequate quality standards, helps reduce costly and disruptive unscheduled downtime.

You want to ensure that your scheduled downtime—by which I mean shutdowns and routine scheduled service outages—is effective and efficient. It should never be cut short in an attempt to improve availability.

People often try to reduce scheduled downtime to increase the availability of equipment. However, you must first control unscheduled downtime before trying to reduce scheduled downtime.

To determine whether a site should reduce scheduled downtime, I look at the amount of scheduled downtime, unscheduled downtime, and the target availability. I want to see how much scheduled downtime there is and how this compares to our benchmarks for mobile and fixed plant. If the amount of scheduled downtime is about right but the unscheduled downtime is excessive, then it is best to target these unscheduled events first.

Ninety-nine times out of one hundred, you must first correct the execution effectiveness or quality standards of unscheduled downtime before you can reduce scheduled downtime. However, there are sites that are doing the basics well and achieving low levels of unscheduled downtime. It is also true that you can work on reducing scheduled and unscheduled at the same time, although you must recognise that they require different corrective actions.

If your site is doing quality work with around 60 percent scheduled downtime for mobile equipment and 70 percent or higher for fixed plant, then you have the opportunity to optimise the scheduled downtime. These solutions must also allow for the same amount of work to be completed at the same quality standards but with less scheduled equipment downtime.

WHAT IS THE CAUSE?

Excessive scheduled downtime typically results from problems with planning and scheduling. We invented planning and scheduling to make our scheduled downtime events more efficient and to give our workers the information they need to do a better job. Planning and scheduling can have some effect on unscheduled downtime, but only if execution standards are high. The goal should be to get more work done efficiently in a smaller amount of scheduled downtime with processes that facilitate repeatability and precision.

Scheduled downtime events need to be planned in detail and should include an assessment of what could go wrong with each task. We then have the opportunity to identify how to avoid those potential delays or at least have a contingency. When you analyse the challenges that disrupt scheduled downtime, you repeatedly find the same issues—most of which can be avoided.

Here are some problems that lead to excessive scheduled downtime and how to solve them.

THERE IS NO ADEQUATE SCHEDULED-DOWNTIME STRATEGY FOR THE LIFE CYCLE OF THE MACHINE

When a maintenance team develops a strategy for a machine, the first thing it must do is look at the required maintenance tasks to be performed over the life cycle of the machine. Then the team must look at how it can fit the tasks into a repeatable sequence of scheduled-downtime events and what the resulting scheduled downtime of the machine will be across its life cycle.

When we start a Bluefield project on a site that is not achieving its availability targets, we almost always find that there is no formal scheduled-downtime strategy. People generally know when the machines are shut down, but there has been no attempt to control the amount of scheduled downtime in a logical manner that allows sufficient time to complete the required work and optimise the downtime. When there is no routine downtime strategy, the site will often take scheduled windows in an ad hoc manner, which will lead to unnecessary downtime.

An example of a scheduled-downtime strategy for a mining excavator is shown below. To successfully implement a strategy like this, both maintenance and operations teams must agree to it. This agreement allows maintenance and operations to optimise the downtime over the longer term. This scheduled-downtime strategy also forms the core part of the asset management plan that we document for each machine type.

Scheduled Downtime Strategy Example

Task	Frequency	Duration	Responsible	Timing
Pre-Start Inspection	Each shift	15 mins	Operations	Shift change
Daily Mechanical Inspection	Daily	30 mins	Maintenance	Refueling
In-Pit Machine Washing/Track Cleaning	14 days	4 hrs	Operations/ Maintenance	Pre-service
PM Service, Mechanical Checks, and General Scheduled Repair (using an 8-step fixed time equalised service sheet regime)	14 days	12 hrs	Maintenance	Scheduled Service
Change Bucket GET	1,000 hrs	4 hrs	Maintenance	Scheduled Service
Undercarriage Inspection	3 months	2 hrs	Maintenance	Scheduled Service
Fire-Suppression System Service/Certification	6 months	6 hrs	Maintenance	Scheduled Service
6,500 Hr Shutdown	6,500 hrs	80 hrs	Maintenance	Scheduled Shutdown
Bucket Change	8,000 hrs	12 hrs	Maintenance	Scheduled on Condition
10,000 Hr Shutdown	10,000 hrs	48 hrs	Maintenance	Scheduled Service
13,000 Hr Shutdown	13,000 hrs	240 hrs	Maintenance	Scheduled Service
26,000 Hr Shutdown	24,000 hrs	300 hrs	Maintenance	Scheduled Service

Once the downtime windows are scheduled in the weekly planning process, sites can't delay or wait for an opportune window in the production schedule to do the work. If you wait for a break in production to do scheduled maintenance, the required resources are often busy on other tasks and mechanics can't maximise their work on the idle machine. Instead, managers must provide access to the plant in a scheduled manner if they want the maintenance crew to stay on top of developing defects. They must follow a logical, scheduled downtime strategy.

DEFECTS ARE NOT RAISED IN THE SYSTEM WITH SUFFICIENT LEAD TIME

Everything that could cause a machine to break down before the next scheduled service must be fixed before it's put back to work. However, not every problem needs to be repaired on the spot. Instead, some defects should be raised in the system and scheduled for the next event when the parts and labour required to execute the job can be planned and scheduled. This is more efficient than doing emergency repairs.

For example, a worker notices mechanical wear on a hydraulic hose. He thinks he needs to replace it, but he doesn't have a spare hose. So he goes to the warehouse and gets one. Then he must find time to replace it within the scheduled downtime event.

However, if the hose wear is so bad that it needs to be

replaced immediately, then the wear should have been noticed and raised in the system during the previous service when the problem was less severe. By waiting until something had to be done, the maintenance team allowed an extended downtime event. That's a problem that could have been corrected if the team was having daily discussions about quality at its morning meetings—an idea we discussed in chapter 4.

Maintenance teams also must be clear about tolerable defects. Not everything has to be fixed on the spot. Remember how the airline industry has detailed limits on such things as drips per minute? Some leaks are tolerable and can be fixed at the next scheduled maintenance, but you have to be aware of the leak in the first place. The mining industry needs to follow the example of the airline industry and ensure all defects are raised and tracked so there is sufficient lead time for the repair to be planned and scheduled.

MAINTENANCE PROFESSIONALS THINK THEY HAVE TO RUN PRESERVICE INSPECTIONS

Preservice inspections, which look for equipment defects three or four days before the equipment comes in for scheduled service, are a waste of time. This strategy does not allow people to plan ahead. What's more, operators perform daily inspections on their machines, so a preservice inspection is either redundant or an admission that operator inspections are not being done correctly. The preservice inspection adds scheduled downtime that isn't necessary.

PEOPLE BELIEVE THERE IS A PERFECT PLAN

When I was a planner, each week we discussed jobs that were scheduled but for some reason did not get completed. We were targeting 90 percent schedule completion. The supervisor would often blame planning (me) when we fell below that level. I questioned the supervisor and held him accountable for executing the plan. After we discussed it, though, most of the blame would come back on me. I didn't feel good causing problems for the supervisor.

I eventually realised that there is no perfect plan. Planning is about continuous feedback and improvement. Every time the fault came to me, I put in place a solution to that problem so it wouldn't happen again, and I would communicate with the supervisors so they would understand what I had done. The planning improved and so did the understanding of the plan by the supervisors.

As a planner, I learned the importance of overcommunicating. For example, we often experienced delays during scheduled service days due to other teams trying to work in the same area. To address this, I introduced GANTT charts for our scheduled service days showing what job had to be done at what time. I also used a color-coded system to show critical requirements, such as when power would be turned off for the high-voltage electrical inspections and when the overhead crane was in use.

Despite this detail, I continued to find mechanical jobs that were not completed because the power was turned off. At our weekly meeting, I would ask, "How did this occur? Did they follow the GANTT chart?" The response I got from the step-up supervisor was, "What's a GANTT chart?"

This is why I say it is crucial to communicate more than you think is necessary so everyone understands the details. There is no perfect plan, but good communication between the planner and supervisor can improve and become very good.

Doc Palmer wrote a tongue-in-cheek article titled "Best ways to kill planning, scheduling." The first step is to believe that there is a perfect plan. The second step is to link planning and scheduling performance metrics to people's pay packets. This second step ensures all metrics will be met, but it doesn't necessarily improve planning and scheduling.

Schedule compliance shows how long the equipment was

down compared to the schedule. Schedule completion indicates how much work was completed compared to the schedule. You need to use these metrics as indicators of performance. When schedule compliance and completion metrics are not meeting targets, you must find and correct the real reasons for the delays. But when schedule compliance and completion metrics are linked to people's pay packets, you find that the metrics are met, but the maintenance is not efficient. It is common to see work orders opened and closed just to indicate to the key performance indicator (KPI) system that the work got done. However, it is easy to achieve good metrics by eliminating the real problems experienced during scheduled downtime events, and this drives real improvement in maintenance efficiency.

Palmer's comments about having a perfect plan reflect my experience. Planners who believe plans should be flawless will assume all failures are their fault and will attempt to fix it themselves. However, planners should be planning for the future, not working on problems in the present. When supervisors experience problems during scheduled downtime, they must solve it themselves. There still must be at least a weekly improvement cycle where supervisors and planners discuss these problems and implement actions or mutual agreements to address them. Planners and supervisors must discuss detailed reasons for the delays. I prefer to discuss these problems every day while they are fresh in everyone's mind. It's hard to remember what you did two days ago, let alone all the detailed problems that occurred a week before. However, if people must follow a weekly review schedule, a maintenance-delay analysis can help them capture what occurred, so the team can implement effective solutions to avoid future problems.

When I was a planner, the supervisor and I agreed that my role was to plan for the future scheduled period, not the current one. The supervisor would deal with whatever problems came up that week. This sort of mutual working agreement is a crucial solution.

PLANNING AND SCHEDULING ARE NOT SUFFICIENTLY DETAILED

If you look at Formula 1 racing, you can see the benefits of planning the details and continually looking for ways to reduce the scheduled downtime or "pit stop." At Bluefield, we teach our clients a simple technique that we call Single Minute Maintenance (SMM).

The SMM process is adapted from the Single Minute Exchange Die (SMED) process, which was developed for the manufacturing industry many years ago. We have adjusted the naming to better suit the mining and resources industry. The process enables planners and teams to significantly reduce their scheduled equipment downtime. By applying a simple process, teams can develop detailed procedures that minimise equipment downtime while maximising work quality and repeatability. Some companies call this standardised work, but the same way of thinking about a job can be applied to the planning process as well.

The process overview is shown below, but more details can be obtained from http://resources.bluefield.com.au/reducing-scheduled-downtime-in-a-sustainable-manner.

Single Minute Maintenance Process

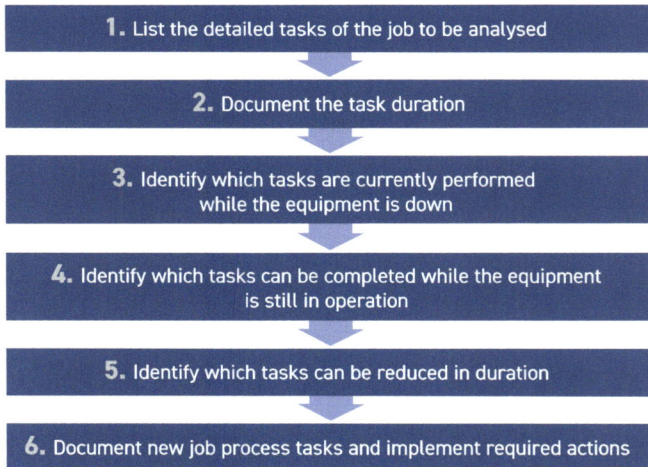

1. List the detailed tasks of the job to be analysed

2. Document the task duration

3. Identify which tasks are currently performed while the equipment is down

4. Identify which tasks can be completed while the equipment is still in operation

5. Identify which tasks can be reduced in duration

6. Document new job process tasks and implement required actions

At Bluefield, we have used this process to reduce a shovel bucket maintenance task duration from fourteen hours downtime to ten hours of downtime. This reduction allows the work to be completed in a standard twelve-hour scheduled service day. On another project, we reduced a mining truck standard service from two people and twelve hours in downtime (twenty-four man-hours) to three people and 4.5 hours of downtime (13.5 man-hours). This powerful process not only reduces scheduled downtime but also improves consistency and repeatability. Many small improvements can deliver a big gain.

CYCLING BETWEEN CENTRALISED AND DECENTRALISED PLANNING (MORE WORK FOR THE MAINTENANCE MANAGER)

Centralised planning means all the planners work on one team with a planning function lead or superintendent. Decentralised planning means the planners are part of the same team, as the supervisor and planners both report to the same superintendent. Many companies have centralised their planning functions, but clients still ask questions about whether centralised planning is better than decentralised planning. In years to come, I'm sure there will be a push to decentralise planning again as the cycle continues.

Centralised Planning vs Decentralised Planning

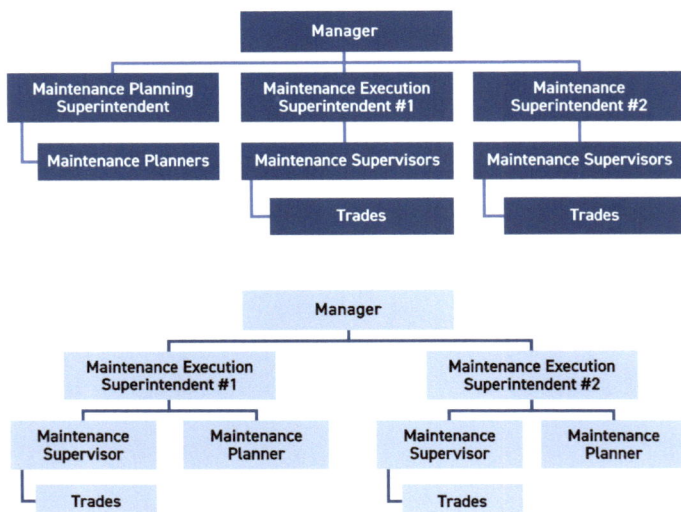

Both structures have problems. Centralised planning typically has problems with communication between the planning and execution functions. If the superintendents of those two functions don't work together and communicate well, the manager has to step in and sort things out. This often creates more work for the manager. When planning is decentralised, you have only one superintendent to go to in order to get better performance from the plant.

In the decentralised model, planners often get roped into helping fix breakdowns. They stop planning and become resources for execution. In addition, decentralised planners often do things differently, causing inconsistent planning approaches across the business. When this occurs, managers want to centralise the planning function to get it in control, which introduces communication problems and removes single-point accountability.

The structure doesn't matter if people commit to making the equipment perform. In fact, restructuring creates a mask; it doesn't solve the problem and often creates confusion. It's more effective if you set up the right communication processes. Whether it's a centralised or decentralised model, planners and supervisors must talk—all the time. They must discuss (in either a meeting or in informal conversations) what went right and what went wrong. There must be formal weekly or daily schedule-review meetings, but informal discussions are equally important.

CHAPTER 6

Budget Overruns and Cost Concerns

In mining, where maintenance represents up to 40 percent of a mine's operating expenses, companies know they need to keep maintenance under control if they want to control costs. However, maintenance expenses are often not managed or understood well.

Earlier in my career, we spent money on work with little thought to the cost. We did not work to a budget. We simply dug up the minerals, sent it to the customer, and made our money.

In the nineties, the mining industry in Australia became tighter, and it became harder to make a profit from those commodities. Mining companies instituted careful cost and expenditure controls, and we couldn't spend over the amount budgeted for maintenance. Then in the mid- to late 2000s, a boom drove up commodity prices and we went back to spending whatever was necessary on people and equipment. In 2012, when everything came crashing down again, mining companies were again forced to bring costs under control. Strict cost controls continue today, and we've learned that sustainable cost reduction comes from first understanding the required maintenance expenditure and then controlling the losses and waste associated with maintenance.

WHAT IS THE CAUSE?

Cost controls can become a rub point between general managers, COOs, CFOs, and maintenance managers. The maintenance manager is often judged for expen- ditures—someone is always saying maintenance costs are too high—but this judgment isn't entirely fair. What measures are executives using to make this judgment? Executives might see expenses go up from one year to the next and think that is a problem. However, equipment life cycles require components to be replaced at certain intervals, so maintenance expenses legitimately fluctuate from year to year. This should be expected, but this fact is misunderstood by many executives. Budgets are not set with equipment life cycles in mind. What's more, execu- tives often don't know how to judge the cost performance of maintenance. They compare sites' performance based on the cost per product tonne, which is important to the bottom line but not useful as a benchmark.

Here, then, are the overarching problems related to budget overruns and what managers can do to address them practically and sustainably.

ACCOUNTANTS AND EXECUTIVE LEADERS DO NOT KNOW HOW TO JUDGE COSTS

Executive leaders often don't know how to judge maintenance costs. They plan on a flat line, but that isn't possible for a piece of mining equipment. The maintenance cost is going to go up and down from year to year according to the machine's normal maintenance schedule. You can smooth the expenditure if you have a large fleet, but that's hard to do on a small site or individual machine basis.

About a year ago, I was at a site where management consultants had analysed the cost of all the machines over the past year. We had improved the fleet availability and there was a surplus of trucks. To improve cost performance, the management consultants proposed shutting down a couple of trucks that had cost the most to maintain in the previous year.

Our engineer argued that the expensive trucks had just

been overhauled and were in good condition. The trucks that should be shut down were those coming to the end of their major component life. Those trucks would have to undergo costly overhauls in the coming two years.

At Bluefield, when we are asked to help a client reduce maintenance costs, the first thing we do is a simple benchmark on where the site is with its costs. We have found two suitable ways to make this comparison. The first is to calculate the maintenance cost per total tonne moved and the second is to compare the maintenance cost to asset replacement value.

Moving tonnes of material is what causes the equipment to wear out and require servicing. However, it is not an apples-to-apples comparison to look at two different sites' maintenance costs per product tonne. One site might have to move a hundred tonnes to get one tonne of product, and another site might only have to move ten tonnes to get the same one tonne. To demonstrate this point, look at the following maintenance cost graphs for several mines. The first shows the maintenance costs per product tonne, and the second shows the maintenance cost per total tonne moved. Clearly, the site that looks the worst on a cost-per-product tonne is nearly the best when looking at cost-per-total tonne moved. This site has much higher strip ratios and needs to move more material to make the same amount of product.

Maintenance Cost/Product Tonne

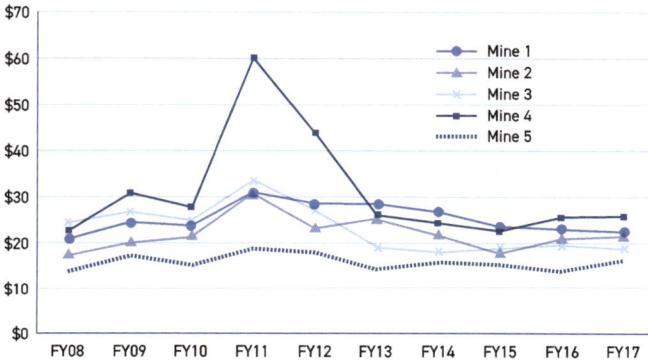

Maintenance Cost/Total Tonne Moved

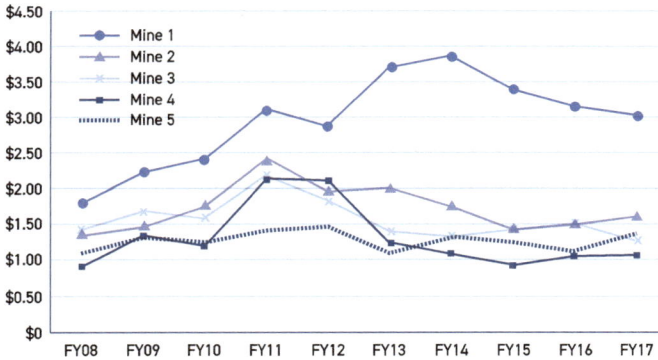

We recently used this benchmark at a site that was forecasting a significant cost overrun. The benchmark showed that there was a clear opportunity to get back to an acceptable level of expenditure, after which everyone was ready to go and find the losses and waste!

Executives can also look at the long-term maintenance cost performance by using a metric called the maintenance cost per asset replacement value (ARV), or sometimes called RAV. This is where you compare actual maintenance costs to the ARV. This metric can reveal when sites are overspending or not spending sufficient money on maintenance. You can underspend on maintenance for a couple of years, but executives need to understand that underspending will eventually lead to greater expenses down the track.

For more information on definitions and how to calculate maintenance cost to ARV, there are several resources on the web. The Society for Maintenance & Reliability Professionals has a good definition and details an approach to calculating the replacement value. I simply use the initial asset value inflated for inflation to today in order to calculate the replacement value. The important thing is not how it is done but to do it consistently. This metric can also be used as a benchmark for sites using similar equipment. We have learned that to use it, you must compare similar equipment or plants. You cannot compare a copper concentrator with a coal wash plant or a non-mining asset, but you can compare large copper concentrators with small copper concentrators. Some data that we have gathered over the years is summarised in the following table.

Site	FY	ARV ($m)	Mtce Cost ($m)	Mtce Cost/ARV
Mine 1	FY-09	136.4	36.4	26.7%
Mine 1	FY-10	137.3	38.6	28.1%
Mine 2	FY-11	104.0	36.1	34.7%
Mine 2	FY-12	106.8	34.0	31.8%
Mine 3	FY-09	606.5	167.2	27.6%
Mine 3	FY-10	612.2	179.8	29.4%
Copper Plant 1	FY-07	273.1	9.2	3.4%
Copper Plant 1	FY-08	295.6	20.1	6.8%
Copper Plant 1	FY-09	306.8	19.4	6.3%
Copper Plant 1	FY-10	322.1	33.7	10.5%
Copper Plant 1	FY-11	331.6	34.2	10.3%
Copper Plant 1	FY-12	340.1	31.0	9.1%
Copper Plant 1	FY-13	348.8	28.2	8.1%
Nickel Plant 1	FY-09	509.6	37.7	7.4%
Nickel Plant 2	FY-09	422.7	24.8	5.9%
Nickel Plant 3	FY-09	146.8	8.7	5.9%
Nickel Plant 4	FY-09	794.8	31.0	3.9%
Nickel Plant 5	FY-09	444.5	22.8	5.1%

Executives can also judge site performance by looking at the cost per hour of the equipment. But this needs to be considered in terms of the machine life cycle. The costs per hour go up and down across the life cycle of the machine.

We utilise a life cycle cost model and cumulative costs to compare maintenance expenses at a machine level. This is also the starting point for finding losses and waste in the machines based on lower-than-expected major component lives.

If you compare your actual costs against the life cycle cost model, you can see where machines are not performing well according to that life cycle plan. Then you can drill into further detail to uncover losses, such as short major-component life or excessive labour.

Cumulative Cost: Baseline vs Actual

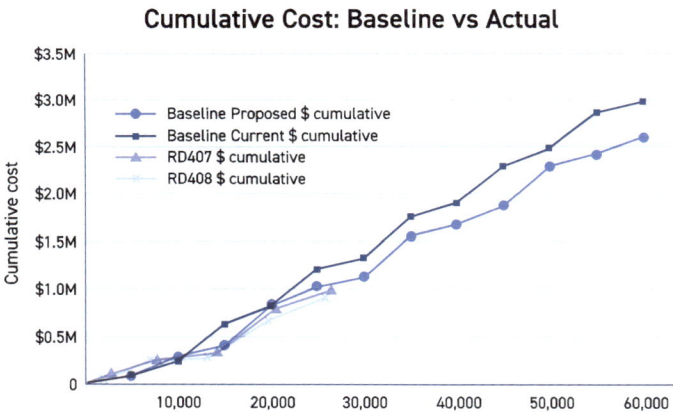

This graphic shows the life cycle cost forecast for a piece of equipment. The cumulative costs go up the longer the machinery is used. The top (blue) line shows a forecast for life cycle costs, and the black line shows a different forecast based on an improved maintenance strategy.

Below them, the data points depict the first couple of years of actual costs over the hours of use. Tracking costs this way is the best method to show the performance of many machines against a baseline and to compare similar machines across a business.

Rather than comparing a machine's maintenance costs from year to year, look at how equipment performs against that life cycle cost model. When you compare costs to the model, you can see where machines are not performing according to the life cycle plan. You can also proactively look at different options for maintaining machines.

You need a proper method for comparing maintenance cost across operations. I recommend the maintenance cost to total tonne moved/processed or maintenance cost to asset replacement value. It is also necessary as part of the plant asset management plan to include a zero-based life cycle cost model. Developing these metrics and cost models does not have to be a major piece of work. Keep it simple and ensure that all stakeholders, from the COO to the supervisors, are aligned with the method you choose.

BUDGETING AND MANAGING A MAINTENANCE BUDGET

Once you have alignment on how to compare cost performance, it is necessary to ensure all levels of management agree to stay within the maintenance budget.

One of our clients at Bluefield was 20 percent over their maintenance budget after the first quarter. We found many areas of loss and waste, but it was also a classic case of one person developing the budget and everyone else spending without regard to that budget. The people spending the money—the supervisors, planners, tradespeople, and superintendents—must know their part of the budget in detail. I recommend the following approach to budgeting:

1. **Involve those accountable for expenditure in the budget-development process.** At many companies, the budget is developed by one or two people in the engineering or planning team. They know the equipment strategy in detail and can efficiently develop a zero-based budget. This is fine for budget development, but these people are not making all of the expenditure decisions. Execution superintendents should also be involved in the budget-development process and take responsibility for any assumptions used to create the budget. This doesn't take a lot of time, and their input will improve the budget and provide ideas for reducing maintenance costs.

2. **Involve those who make expenditure decisions.** Many expenditure decisions are made by planners and supervisors. If these guys have input to the assumptions and desired outcomes, this further deepens their understanding of the budget details. This enables them to work to the budget in a simple

manner, and they always have ideas for maintenance cost savings.

3. **Develop a reference tool or book.** When the budget is finalised, you must have a simple reference tool or book so supervisors, planners, and superintendents responsible for spending decisions can easily determine if an expenditure is in the maintenance budget. If not, there must to be directions for identifying improvements or alternatives so the actual costs are reflected in the maintenance budget each year.

4. **Capture knowledge and continually improve the maintenance budget.** You must record and learn from times when expenditures are outside the budget. Using cumulative life cycle cost management tools can highlight these instances, and reviewing this on a monthly or quarterly basis may reveal further maintenance cost savings.

While 20 percent of the machine components will cause 80 percent of your maintenance costs, reducing and controlling maintenance expenditure requires that all areas are managed. Many small expenditures can quickly become a large amount of money if not controlled.

WORKSHOP ORDER AND DISCIPLINE DO NOT MINIMISE PARTS WASTE

Mining companies lose tens of millions of dollars by

poorly managing parts, allowing components to be lost, damaged, or become obsolete.

Managing parts is simple if you document a process for them. However, it's often difficult to implement the routines across all rosters and between maintenance and supply functions.

Use a Five-S structure to set up a process. There are many resources that help you implement a Five-S program. It's easy to get to the fourth S, but it is essential that you achieve the fifth S and make your system sustainable. Here's how:

- Sort—Clean up the area and throw away all rubbish.
- Set in order—Make sure there is a place for everything and everything is in its place.
- Shine—Focus on maintaining cleanliness.
- Standardise—Establish routines and best practices.
- Sustain—Create a culture of people who care about the work.

Several years ago, I set up a routine managed by the checklist below. We were losing parts and experiencing delays, and I wanted to attack the problem with a logical, orderly process.

Unfortunately, I did not get sufficient engagement from the teams. People took a "tick and flick" approach where they ticked boxes on a checklist and flicked the paper. Consequently, the workshop did not achieve the intended standards.

Daily Workshop Housekeeping Inspection Record

Inspection/Work Area Cleanup to Be Performed from 10:00 a.m. to 10:30 a.m. Daily

Date: _____ Inspected by: _____

Item		Compliance	Observation/Action
1	Walkways clear		
2	Floors clean		
3	Workbenches clean/tidy		
4	Workshop apron clean		
5	Hoses rolled up		
6	Parts washers closed/clean		
7	Portable steps in good condition		
8	Portable steps stored correctly		
9	Lube station area clean		
10	Hydrocarbon storage/leakage		
11	Jacks/safety stands stored correctly		
12	Tyre bay area clean		
13	Signage in place		
14	Tools cleaned and stored		
15	Gas cylinders secured/stored correctly		
16	Boilermaking area clean		
17	Hardstand/park up area		
18	Washpad clean		
19	Washpad traps OK		
20	Washpad Hoses rolled up		
21	Parts to be sent for repair tagged and in correct location for pickup		
22	Parts to be sent for credit tagged, prepared, and in correct location for pickup		
23	Bins emptied		
24	Lunchroom supplies adequate		
25	Library area tidy		

This approach also created more work for me, as the leader of the team, to ensure compliance. I pushed but eventually asked the teams to develop their own standards and procedures. This approach was more successful. The teams agreed to set aside the same time each roster cycle to reorganise the workshop and return all parts to the designated areas.

Workplace stores, squirrel stores, or direct-purchase storage areas for each machine also contribute a lot of parts losses. The warehouse is much better than the maintenance team at managing parts, so I recommend using the systems that are in place and making them effective rather than creating workplace stores.

In some of the site reviews Bluefield has conducted, we found more than $50,000 worth of parts left over from completed jobs just for one machine and millions of dollars' worth of parts across large sites that were ordered but never used. We found at one small site more than $400,000 in ground-engaging tools that were never used and became obsolete due to changes in the equipment.

Teams must develop an attitude that workshops must be clean and organised. A maintenance manager I worked with in Chile said every person on the team, including supervisors, must stop work at the same time every day and organise the workshop. Teams must develop these standards, and the standards need to be documented and embraced by all team members.

PLANNING AND SCHEDULING DOES NOT MINIMISE LABOUR WASTE

We touched on this when we talked in the previous chapter about reducing scheduled downtime by eliminating maintenance delays. It is important to create an environment where scheduled maintenance is completed efficiently. A time-on-tools metric to measure workers' efficiency is difficult to use on an ongoing basis, and you must do a time-in-motion study to determine what the time on tools is. So what's the best way to create more efficient scheduled maintenance?

First, I assess the labour required, based on the equipment and required maintenance. Once the maintenance manager and general manager agree on the full-time equivalent (FTE) resources that are needed, you can monitor the FTE resources actually used to complete the work. This allows you to track performance against your cost benchmark.

Then you can apply the processes that we discussed previously to capture maintenance delays and make planning and scheduling improvements, as well as implementing the principles of SMM.

As I mentioned in the maintenance improvement model, I have implemented the maintenance-delay reporting process. This works only if everyone agrees on what constitutes a delay.

Every moment people are standing around is wasted time and needs to be reported. If people accept it as routine and don't talk about it or report it, there is no chance to continually improve. This impacts both workforce efficiency and scheduled downtime. Inefficient labour increases costs.

RELIABILITY-CENTRED MAINTENANCE (RCM) IS MISUSED

This issue could be put in any of the major problem categories. I include it with cost because people often spend a lot of money developing maintenance strategies using RCM methodologies but don't get any value out of it.

RCM is a brilliant methodology for thinking about how machines fail. John Moubray wrote the fantastic book *RCM II: Reliability-Centered Maintenance*, and many companies help to facilitate RCM workshops in the mining space.

However, it's expensive to bring in companies to facilitate the process and to send your people to the workshops. Companies that adopt RCM—or even a modified RCM approach, such as a preventive maintenance optimisation methodology—do this through a project-oriented approach designed to develop the perfect maintenance tactics. The goal is optimisation, but from my experience, the project-oriented approach has been ineffective in mining.

The methodology is a great way to think about how machines fail and how to manage those failures. Everyone should think this way when trying to improve machine performance and reliability. However, the industry is littered with mines that spent a fortune on RCM maintenance strategies that never paid off. This does not mean a well-implemented, practical RCM or a Failure Mode and Effects Analysis (FMEA)-based project cannot add value, but it must be implemented correctly.

For example, at one job site, we had a pressure relief valve fail to function. With any type of pressure tank, over-pressurisation can cause the tank to rupture. This is prevented by the over-pressure valve, which opens and relieves the pressure when the tank pressure gets above a set level.

I found out about this event when the superintendent there, who had years of experience, told me he wanted to change the maintenance strategy for that valve. He reasoned that since the valve's expected lifespan was 6,000 hours, we should replace it at 4,000 hours of use.

That may sound sensible to many people. The problem was that the actual failure mode was not time-based; the failure was due to being clogged with some sort of material buildup that occurred at a random time interval. This can happen after 5,000 hours, or it can happen two weeks after installation for all we knew, because no one could see it or know it failed until the valve stopped working. So replacing it at 4,000 hours didn't make any sense.

The only way to ensure the valve was still working was to test its function at an established frequency, such as every three months. So, for this situation, we got another valve that could be tested and implemented a testing regimen.

This example is not uncommon. Even experienced people can waste time and parts when they don't have an RCM mindset. Rather than using RCM on a project basis, we must think about managing failure modes and solving reliability problems day in and day out.

Dick Pettigrew was an early adopter of John Moubray's work and was mentioned in the *RCM II* acknowledgements. Dick helped me learn and understand this important aspect of RCM. Dick used RCM in the US chemical industry as an ongoing way of examining how things fail and how to continually improve them.

CHAPTER 7

People Problems

This chapter addresses people-related items that don't fit into the other categories. You could argue that every problem in this book involves people—people aren't motivated, they're not committed, or leadership doesn't care. You could say all of the difficulties in this book come back to people. In the end, it's not the machines but the people who make the most difference in the workplace. This chapter will deal mostly with issues of relationships and how people work together.

WHAT IS THE CAUSE?

We have already discussed how important open and honest communication is. Poor communication leads to significant problems and can create a workplace where people don't help each other out. You can't expect someone to go the extra mile for you if you don't do the same. I have worked in places where the manager has to make all of the decisions instead of the frontline workers figuring it out themselves. This is a divisive approach; people work against each other rather than as one team. Operations like this typically have low morale and high turnover, which leads to many of the problems we've discussed previously.

People in cross-functions often don't recognise that they are all working on the same team in one business. As a manager, it bothered me that we couldn't all just work together as one company. Unfortunately, many people seem to need someone within the business who is the bad guy. I call it the internal bad guy syndrome. If it's an operation where everyone works well together, they will identify someone from corporate as the bad guy. If it's a company where everyone gets along, they will find an industry regulator to treat as the bad guy.

This mentality isn't productive, and the only way to change it is if leadership, whether it's on site or in a corporate office, models good communication and goes the extra mile for others.

Here are some more specific people problems and how they can be overcome.

MAINTENANCE MANAGERS TALK ABOUT MORE PEOPLE AND MONEY RATHER THAN BUSINESS PROFIT

General managers often have problems with maintenance managers. Maintenance managers always want more people and more money, particularly if they are in a reactive environment. But general managers don't want to hear that. General managers need to achieve higher profit, and requests for more money threaten that goal.

The simple solution for maintenance managers is to talk to their general managers about profit or the bottom line outcomes that the general manager must deliver. Maintenance managers need to show how their requests for people and money will deliver the profit the general managers want to see. It is not enough for a maintenance manager to tell the general manager, "I'm going to spend this money, and we will get more uptime in the plant throughout." Maintenance managers have to indicate how the money will increase profits, and then they must deliver on that promise.

I have often been caught in the middle of maintenance manager and general manager relationship breakdowns, and if the conflict cannot be resolved, the maintenance manager usually leaves for another job. The general manager is also impacted, because they don't achieve their business goals.

Bluefield once sent a temporary maintenance manager to one of our clients' sites. Our employee had never been a maintenance manager before and did not know the general manager. Before he took the position, I coached him: "Whatever you do, do not talk to the general manager about more people or more money to make improvements. Talk to him about profit. Talk to him about what he wants to hear. Talk to him about how you can help his business perform better. Understand what he needs to achieve."

This guy followed my advice and had a great relationship with the general manager. He received the money and people he needed by showing how additional resources would help the bottom line. More importantly, he delivered on his promises. By focusing on what the general manager needed to deliver, he developed a trust that made their relationship fruitful.

The key is to demonstrate how additional resources will benefit profit. If you know you need to do a shutdown and want to bring in a specialist to reduce downtime, you would articulate the plan to the general manager like this: "The plan is for this machine to be down for three weeks. This is a critical machine in the business. Every hour lost with that machine is lost product and lost tonnes. I want to reduce this downtime from three weeks to two, but I need to spend $150,000 on a specialist project manager to help me do that. The payback will be in the millions."

When you provide a clear business case like that, the general manager will likely see the wisdom and grant your request, especially if you have a proven record of delivering the outcomes you promise.

LEADERSHIP EXPECTS SUPERVISORS TO MAKE ALL OF THE DECISIONS

Increasingly, people in the business think the supervisor needs to make all of the decisions. This leads to the mentality that when anything goes wrong, it's the supervisor's fault. As a supervisor, you generally have twenty to thirty people on your team, and if those people come to you all day to solve their problems, you'd never get your own work done.

Mine leadership must empower tradies to make their own decisions and take responsibility for their actions. In this scenario, the supervisor becomes the coach of the team while the other team members self-reliantly do their work. Good supervisors do not try to do everything for the team. Instead, they create high-functioning teams under them.

When I was a supervisor, I learned to entrust people. For example, one guy came to me every day to sign a piece of paper allowing the guy to get consumables from the warehouse. One day I taught him how to request things from the warehouse himself. He became happier because he could now function in his job without being held up by me, and I got more time to focus on more important things.

Another example involved the machine operators, who called me every time they had a breakdown. It took a while, but I got them into the habit of directly calling the tradespeople who would respond and fix the machine instead of calling me and asking me to call the tradies. This empowered the operators as well as the tradespeople, who now made more decisions for themselves. I often see work-management processes dictate that the supervisor makes all of the decisions rather than the tradies who know more about the problem.

A supervisor cannot and should not be watching the workforce all day. It is essential to have empowered tradespeople to make decisions.

To achieve that environment, the supervisor must coach and empower the tradespeople. If you are a supervisor, every time a tradesperson comes to you with a problem, ask yourself, "Is this a decision I need to be making? How can I empower this guy to address the problem rather than going through me?" These questions generate solutions and improvements.

Teams that have talked about the foundational values

of the maintenance improvement model, which we discussed in chapter 2, will know the right decisions to make. Sometimes we try to solve problems by writing prescriptive procedures, but this is too complicated. If we focus our teams on the values that we have developed, they will make the right decisions when the situation arises. This is a much simpler approach.

EXECUTIVES DO NOT KNOW HOW TO JUDGE THE PERFORMANCE OF MAINTENANCE

I once consulted for a business that had eight different mining operations. I conducted a performance review on each one and found that the last site was the best performer. The workshop was orderly, and the maintenance manager had been there for five years and had the same maintenance superintendents and supervisors during that time. He was open, honest, and direct, and the metrics around scheduled and unscheduled maintenance downtime also reflected the good performance of the plant.

Despite the maintenance manager's excellent work, the general manager didn't care for him. The maintenance manager may have been too direct in his communication. In any case, the maintenance manager eventually left that company for a mining job elsewhere.

This illustrates that maintenance managers are sometimes not judged correctly. In this example, the availability of the equipment was good, but there are also many examples where maintenance is judged on availability alone, regardless of whether the downtime was scheduled or unscheduled. As discussed in the previous section, there is also a disconnect on comparing maintenance cost-performance.

If a machine goes down and stops producing, general managers know because their people stop working. Maintenance managers, on the other hand, won't instantly know when a breakdown occurs, because their team is on the ground fault-finding. What usually happens next is the general manager asks the maintenance manager when the machine will be back up, and if the maintenance manager doesn't know, he's incorrectly judged as being unaware of what is happening—even though the supervisors and tradespeople haven't found the problem yet.

The solution is for the general manager and the maintenance manager to have an open and honest discussion about how the maintenance team and performance should be judged. The general manager must understand that the maintenance manager won't immediately know the issue in a breakdown and instead should be judged on scheduled downtime, unscheduled downtime, and life cycle performance of the plant. They should be judged on their trends and whether they are eliminating repeat failures. What is done today in maintenance will show in the results six to twelve months down the track.

CHAPTER 8

Project Transition to Operations

My company's name, Bluefield, came from two other common mining terms—greenfield and brownfield. A greenfield project is when a mining company develops a new deposit of minerals or other commodity at a site where no one has mined before. The company must purchase equipment and build a new mining plant to get more production from this new equipment. A brownfield project is when a company expands operations at an existing site. This involves buying more assets and equipment or building another plant—anything you might need to increase production. A Bluefield project is when companies get more from their existing assets or get more than expected from the new assets they are already developing.

This chapter deals with the last key problem area in maintenance, which occurs when greenfield and brownfield projects transition equipment to the operations and maintenance teams.

Whatever the project, the maintenance team often is not adequately involved in selecting and purchasing the new equipment. In a large project, a dedicated project team handles this job. Even in smaller projects, the maintenance guys may not work with the project team in selecting and purchasing the equipment. The project people are professionals but often don't have any background in maintenance or operations and are always driven by different deliverables. As a result, they often miss what must be considered from a maintenance standpoint.

Several years ago, Bluefield was called to a mine in Botswana. The mine was sitting on top of a decent deposit, and even though it was only about two years into operations, the mine's equipment was showing heavy wear and availability was declining.

We found that the maintenance guys had inadequate tools and facilities. They had no workshop to maintain the equipment. They had no way to get their oils, lubricants, and other fluids into the machines in a clean manner. They couldn't store their lubricants out of the weather. They had to pour oil out of forty-four-gallon drums into smaller containers to transfer it to the equipment. It was difficult to keep the dirt and water out of these lubricants, and this caused premature wear on the machinery.

We worked at the site for several months and put in many corrective actions. The mine began to make money.

However, about a year later, the company ran out of money. The equipment quickly began to break down. As one machine broke down, pressure to get more production from the remaining machines increased. As those remaining neglected machines began to fail, production fell to unsustainable levels. Despite the generous deposit, the company that owned the mine eventually went out of business.

WHAT IS THE CAUSE?

Operational readiness, including maintenance readiness, has been around for some time in the mining industry. Not all industry experts agree on how to measure maintenance readiness, but here are the minimum requirements:

- Solid life cycle plans for all equipment
- Adequate spares
- Facilities, tooling, and systems for working
- People with knowledge of the equipment
- A plant designed for maintainability, reliability, and standardised equipment

Problems arise when maintenance is not involved early in plant design or equipment selection. For example, the maintenance team might realise after a new plant starts that workers can't access a pump or that there is no overhead lifting device for removing the pump when it needs to be replaced. I've seen workshops where the project design team didn't leave enough space for trucks to enter, let alone be repaired.

Problems like these can be avoided if the maintenance team has early input. The project team has a deadline and a budget, too, so design input has to come early, when changes to the design are more costly to make. Project teams don't like to admit to making mistakes, so they will defend their design when maintenance comes in after the fact and wants to change it. This creates disagreement and arguments. These problems are less likely when the design team includes maintenance personnel from the initial planning phase.

PROJECTS DO NOT CONSIDER MAINTENANCE READINESS

Most larger mining companies consider maintenance readiness during the design and project phases, but too often maintenance is an afterthought. Mining companies must think about maintenance readiness from day one; if there is no accountability or incentive to plan properly for maintenance, project teams will care only about building the plant as quickly and cheaply as they can.

The project team needs an adequate budget. If money isn't provided for maintenance readiness, the project team won't consider it. Some companies have a big bucket of money to cover whatever arises. Others cover every detail as a line item in their budget, so if there is no line item, there's no money for things you forgot.

Regardless of your budgeting approach, if you're a mine executive planning a project, you must consider maintenance readiness in the earliest, prefeasibility phases of the project. Since the design isn't finished at this stage, planners won't know what equipment parts and spares will cost, so they can't plan exactly for maintenance. However, a good rule of thumb is to include 1 percent of the total estimated project capital cost for maintenance readiness, with an additional 5 percent of the physical equipment cost allocated for purchasing spares. We have tested these figures many times, and they hold true for most projects that are over $100 million.

In addition to allocating money for maintenance readiness early in the project, it is essential that the team designs for maintainability and reliability.

Companies usually have some form of design-review process, but many changes requested during the design-review workshops often are not included in the final design. This also frustrates the operations people because they will have to make modifications to the plant.

A solution to this problem is to create a set of principles and rules for the engineering teams. The teams then must demonstrate at the review workshops that they have followed these principles. This means they are designing with this intent from the start, and much of the rework can be avoided. This list can be used as an example of how to structure those principles:

- Commonality of components shall be maintained throughout the site, including existing operations wherever practical.
- Unproven design or equipment shall not be incorporated unless accepted by the owner.
- The design shall facilitate ease of access for handling, transportation, installation, adjustment, control, maintenance, and repair.
- Access to operating and maintainable equipment is to be by permanent access platforms and stairway only. Temporary access platforms (scaffolding and elevated work platforms) can only be used subject to the principal's approval.
- Access is available for all lubrication and isolation points.

The design team must demonstrate how it has met the intent of the principles and rules.

Whatever principles your company develops should focus on preventing problems you've faced in the past. A list of specific problems experienced is also a valuable resource for the design team.

The financial success of any greenfield or brownfield project relies on the plant and production ramping up to nameplate in the shortest time possible. Getting the plant to nameplate requires it to be designed for maintainability and reliability, and the maintenance and operations teams set up for success from the start. Again, this requires alignment across teams, including the projects and operations teams. Setting aside 1 percent of the project capital expenditure for maintenance readiness and aligning the teams may not be perfect, but it pays huge dividends.

MAINTENANCE PLANS ARE NOT OWNED BY THE TEAM

Even if the project team considers maintenance readiness and develops all the necessary maintenance plans and information, the maintenance team must own the information, or it will not deliver. I reviewed a copper mine project in Chile many years ago that had put a lot of money and effort into maintenance readiness for the plant. They had developed significant, detailed maintenance information, but the maintenance team hadn't been sufficiently involved. As a result, during the first year of operation, the maintenance team reworked the maintenance plans for the plant because those plans were not designed the way the maintenance team would have done it.

If you develop a maintenance checklist or life cycle plan for equipment without consulting the maintenance team, they may dismiss your work. The maintenance team may think it's overly complicated, they may not fully understand it, or they may find it different from what they are used to. If they find one error, they can lose confidence in the whole program.

Most people think about operational and maintenance readiness up until the plant is commissioned and handed over to operations. We have found that it is necessary to continue thinking about maintenance readiness through the first year or two. It's wise to include plans for this period in the maintenance-readiness budget.

It's nearly impossible to develop maintenance documentation during a project construction phase without creating errors. Even for mobile mining equipment, the vendor information is often unclear or wrong.

During commissioning and start-up, it is also difficult to bring the new maintenance team up to speed with the plant and information that has been developed. Instead of trying to make all of the information perfect from day one, allow time and resources after commissioning to improve the maintenance program with input from the end users. This time allows for deep understanding and to correct errors that will exist.

At Bluefield, we did the maintenance readiness for a greenfield coal-export terminal. When we started, there was no access to parts of the plant that were being constructed, and there were many examples of the "as built" being different from the engineering documentation. We planned to involve the maintainers in the validation of the preventive maintenance (PM) program that was developed, but when the maintenance team members were hired, they were busy getting inducted and brought on board. They didn't have time to review the plans or develop an understanding of the plant. We learned that it is better to leave the validation and final development of the PM documentation until after commissioning and during the first year of operations.

We also had similar learning for a mobile-equipment project. The information about the service kit from the original equipment manufacturer (OEM) was found to be incorrect after the parts started arriving on site. We went back after twelve months and found the service kits had too many filters included. The filters were being wasted.

There is no perfect handbook for operational readiness. We have done a lot of these projects and we've seen people try to do everything up front. They put so much money and effort into trying to get it perfect up front, only to fail or fall short. The best approach is to continue the maintenance-readiness process for the first twelve months or longer. Just relax somewhat. Spend less money up front and dedicate resources for improving these plans so maintenance crews can be involved and take ownership.

One of my colleagues, one of the most experienced maintenance-readiness managers for copper concentrators in the world after completing three projects on some of the largest concentrators, agrees with this approach.

When Bluefield engages in these types of projects, we set them up from the start with these expectations. This way, work delivers on its intent and avoids waste.

CONCLUSION

I started in a practical role in the line, worked my way up the ladder, and took on all the different roles you can have in maintenance. I learned all the maintenance and reliability theories, but in the end, I came back to what works.

KEEP THINGS SIMPLE. DEVELOP A CULTURE OF CARE, OWNERSHIP, AND A DESIRE FOR IMPROVEMENT. IF YOU SIMPLIFY, PEOPLE WILL ALIGN AND SUCCEED. PEOPLE WILL SORT OUT THE DETAILS.

When I started Bluefield, I was following a long-held dream of opening a business and working for myself. After eight years, I'm still doing it and loving it. People pay us to do the things we're passionate about: helping operations to become leaner, better organised, and more profitable. I get satisfaction out of the work because I get to deliver something of value.

I have the same passion about preventing machines from failing as I had on day one, when I was an apprentice electrician on the workshop floor in Moranbah, Australia, my hometown back then. Over the years, I've developed a greater goal: to help the mining industry be recognised as excellent at maintenance. I want our maintenance people to become the best in the world at what they do. I want to give them back the time they need to be innovative so they're not stressed or working in a reactive environment the way many are today.

I also want to give back and help tradesmen like myself become engineers. I think it's important to enable and support them in that path. Our organisation has supported them with finances and time to study, and we will continue to do so.

We also want engineers to understand the practical side of maintenance. We want to help them better understand the people side of things and the hands-on strategies they can't get at university. We want them to design and devise new ways to improve machinery, but we also want them to appreciate practical maintenance and how to achieve it simply and dependably. Problems can't always be designed out.

I also want to help youth develop a positive internal dialogue that helps them build self-confidence. When I think back to when I was younger, before I was exposed to this way of thinking through books, I didn't think I could do anything. Reading has shown me how to believe in myself, and I'm still sometimes surprised by what I've been able to achieve.

Still, I want to do more. At Bluefield, we have great aspirations for the industry. We want to continue to provide specialised services, knowledge, technical support, and systems to improve maintenance. We're working on a mobile phone app that people can use to locate faults on machines and help improve our collective knowledge of those machines. We're always looking for ways to have a bigger impact on the industry.

Bluefield has grown to more than fifty employees in Australia. We have six in our Chile office and a guy in Finland who is an important part of our family. We have worked on all continents and are continuing to grow. We've just come through five years of the hardest downturn the mining industry has seen in a very long time, which came after the biggest boom. Despite the hardships, we have managed to grow our team.

Now that the mining industry is turning around and commodity prices are going up, we're investing in other systems and growing our team even more. We want to be a part of the mining industry's success. We want to be part of the solution to the industry's problems. My partners and I have been in the mining industry our whole lives. Mining is in our blood. We want to see the mining industry succeed even more, especially in our backyard—Australia. I'm disappointed when I see better performance outside of Australia. We have the best education, the best training, and the ability to get more experience than most other countries in the world. But we must get the basics right. I want to see that change and be a part of that solution. We must continue to strive for the latest technology and improved systems, but without doing the basics well, these advances cannot be successful.

This book is a reflection of the passion I feel.

There is no one way to solve everything. My goal with this book is to share the experiences I have had that helped me solve problems. The ideas in this book work, because I've seen them work—they've worked for me—but I am not suggesting these ideas are a prescription that you should follow to the letter. Instead, think about this book simply as the story about how one guy dealt with these problems—the same problems you're probably facing. If my approach resonates with you, put these ideas into practice.

Implementing a transformation process focused on developing a culture of ownership and proactive maintenance may seem daunting, but our experience is that it is not that hard. It just takes the team to see the reality, accept responsibility for performance, and implement the rituals that develop the culture. For more on this idea, see https://bluefield.com.au/bluefield-transformation/.

Nine times out of ten, the answer to any problem will come back to the value bricks in my Maintenance Improvement Model. Remember, the most important brick in that foundation is the desire and determination to improve. If you have that and your team has that, you will find answers to your problems. This book is primarily about people and how to think about problems. It addresses the core of the issues instead of working on symptoms. When there is alignment, agreement, and mutual accountability, people will find their own way to solve little technical things.

I'd also like to recommend a few books that have helped guide my thinking on simplifying mining maintenance. I've already mentioned one book, *RCM II: Reliability-*

Centered Maintenance by John Moubray, as essential reading for anyone interested in a practical approach to maintenance, whatever industry you work in. I highly recommend *Simplicity* by Edward de Bono, who explains how our rapidly changing technology will make "lateral thinking" and simplified processes even more important in the future. I also liked *The Knowing-Doing Gap* by Jeffrey Pfeffer and Robert I. Sutton. This book shows how companies can turn their understanding of performance improvements into actions that reward them with measurable results. This book is filled with great examples of how to turn your knowledge into action.

I also encourage you to visit my company's website, www.bluefield.com.au, where we have several blog posts that expand on some of the points I've raised in this book. Feel free to drop me an email or read about some of our learning from around the world, which we feel is important to share.

I want you to believe you can have the best-performing maintenance organisation in the world. I don't want you to think it is too hard. It isn't. Don't give up. Find that passion that drives you to improve. You don't need complicated processes and documents. Simplify things and make it easier for people to get their work done. Develop that culture. It all starts with believing you can do it.

About the Author

GERARD WOOD is one of the mining industry's foremost authorities on proper mining equipment maintenance. In his long career, Wood has been all over the world, working his way up from an electrician's apprentice to a maintenance manager with advanced degrees in electrical engineering and business. As managing director for Bluefield AMS, Wood helps the world's largest mining companies keep their machines running with a simple, practical approach that saves money and improves equipment reliability.

www.ingramcontent.com/pod-product-compliance
Lightning Source LLC
Chambersburg PA
CBHW041932220326
41598CB00055BA/25